GWR TWO CYLINDER
4-6-0s and 2-6-0s

GWR
TWO CYLINDER
4-6-0s and 2-6-0s
Rodger Bradley

DAVID & CHARLES
Newton Abbot   London   North Pomfret (VT)

Half-title page. Preserved Manor class 4-6-0 No 7827
Lydham Manor on the Torbay & Dartmouth Railway.
R. E. B. Siviter

Title page spread. Two of the GWR's two-cylinder 4-6-0s,
Manor class No 7820 *Dinmore Manor* and County class
No 1008 *County of Cardigan*, at work in South Devon as
they climb Dainton bank on 4 September 1958 with a
Paddington–Plymouth train. *T. E. Williams*

British Library Cataloguing in Publication Data
Bradley, Rodger
GWR two cylinder 4-6-0s and 2-6-0s.
1. Locomotives—England—History
I. Title
625.2′61′0942   TJ603.4.G72G73
ISBN 0–7153–8894–0

Typeset and printed in Great Britain by
Redwood Burn Limited, Trowbridge, Wiltshire
for David & Charles Publishers plc
Brunel House   Newton Abbot   Devon

Published in the United States of America
by David & Charles Inc
North Pomfret   Vermont 05053   USA

CONTENTS

Mogul on express duty: 43XX 2-6-0 No 5368 heads a
Birmingham–Malvern train at Stourbridge Junction in
the 1930s. *Locomotive & General Railway Photographs*

INTRODUCTION AND ACKNOWLEDGEMENTS

The Great Western Railway whose 150th birthday was celebrated in 1985 has long proved to have attracted the affections of perhaps the majority of railway enthusiasts. An important aspect of the GWR's development has been the singular if not unique manner in which the locomotive design and construction policies have been pursued.

It was in the last years of the 19th Century that the company produced its first two-cylinder 4-6-0, No 36, with two inside cylinders, outside frames, and intended for goods working. It was built under the guidance of William Dean, just four years after the broad gauge was abandoned. Dean was responsible, too, for what may well be considered an early mixed traffic type, the Duke, or Devon class 4-4-0. During Dean's last years in the office of Locomotive Superintendent, his assistant, George Jackson Churchward came almost to have a completely free hand in matters of locomotive design policy, as Dean's health deteriorated. Many of the later designs of 4-4-0 saw increasing evidence of Churchward's influence, most notably in the area of boiler design – even some of the illustrious Dean Singles were treated in later years to taper boilers.

It was after Churchward's promotion into the office of Chief Mechanical Engineer that the real changes began to take place, although in 1902, shortly before Dean's retirement the first prototype two-cylinder 4-6-0 appeared. Churchward's design policies were vastly different from all that had gone before, and heavily influenced by contemporary North American practice, which was visibly austere by comparison with the previously ornate Victoriana. In the area of boiler design, Churchward's ideas, as exemplified by the GWR in the yearly years of this century, affected the thinking of many of his contemporary engineers, and the development of locomotive design down to the British Railways period. In mixed traffic design, before his retirement in 1921

Churchward had presided over the introduction of three important classes of two-cylinder tender locomotive, the Aberdare 2-6-0, the 43XX 2-6-0, and the Saint class 4-6-0. In his original scheme for a series of standard locomotive designs for the GWR, Churchward also included a design for a mixed traffic type with 5ft 8in coupled wheels, which later materialised under C. B. Collett, as the Grange class 4-6-0.

In his turn, Collett continued the policies of Churchward with little change in engineering practice, but produced some of the most popular and widely-known GWR two-cylinder types in his 20-year term of office. The first to appear were the Halls, which were based on extensive modifications to the Saint class 4-6-0 No 2925 *Saint Martin*, and resulted in the most numerous mixed traffic type on the company's books. Grange and Manor types followed in 1936 and 1938. Although the former proved to be a very able and popular machine, the Manor was a very poor performer initially, and required the results of extensive testing by British Railways before achieving its true potential. These latter two designs included components from withdrawn 43XX 2-6-0s, although the company's plan to convert all the remaining 43XX types was cut short by the onset of war in 1939.

Major engineering changes arrived with F. W. Hawksworth, who took over as chief mechanical engineer in 1941, and saw the company's transition into the Western Region of British Railways in 1948. With Hawksworth came the Modified Halls (No. 6959 onwards), and the 1000 County class with new self-trimming tender, higher superheat, and changes in cylinder construction. The elimination of Churchward's composite bar/plate frame assembly for locomotive and leading truck brought the company more into line with contemporary practice. Re-draughting of the Churchward designs, with the elimination of blastpipe

Hall class 4-6-0 No 5929 *Hanham Hall* passes Castle Cary on 24 April 1961 with an up parcels train.
G. A. Richardson

jumper ring, and altered front end proportions improved the two-cylinder breed substantially in the 1950s. Overall, in 50 years, the changes that were made to Churchward's original designs were not extensive, with no frequent rebuilding into a variety of sub-classes of locomotive. In many aspects of engineering design Churchward adopted simple but very effective practices, which not only resulted in the continuation of many of those ideas through successive chief mechanical engineers, but which provided a range of very attractive locomotive types.

I am indebted to a number of people who have spared their time to help in the preparation of this book. In particular I would like to thank Colin Jacks, W. E. Peto and the Great Western Society Ltd, Jim Colley, *The Engineer* for permission to use certain diagrams, *Engineering*, and most of all my wife Pat for her support and encouragement.

CHAPTER ONE

THE FINAL YEARS OF WILLIAM DEAN

William Dean was Locomotive, Carriage & Wagon Superintendent of the Great Western Railway between 1877 and 1902, responsible for the introduction of many famous locomotive types. Towards the end of Dean's career at Swindon his eventual successor, George Jackson Churchward, began to introduce quite radical changes, completing his experimental work while still in a subordinate capacity. The GWR board of directors had given Churchward full authority to carry out this work, although Dean was still in overall command. The reason for this subtle and gradual shifting of power between the two men was mainly Dean's deteriorating health, until he was finally persuaded to retire in the summer of 1902.

Although the general design of future GWR motive power and its development during the 20th Century was established after Dean's retirement, some early building blocks for later successful two-cylinder designs had already been laid. The period is also noteworthy for the first introduction of the 4-6-0 wheel arrangement on the GWR, and some of the most famous and successful designs of locomotive, from the Dean–Churchward collaboration.

During the 1890s there were no mixed traffic locomotives as such in the company's stock although there were a number of excellent passenger types, conventional goods engines, and several interesting prototypes.

The GWR's arrangements for working the heavy West of England services at this time could be said to have given rise to introduction of a mixed traffic type. Typical arrangements for running expresses from London to the West Country involved the use of two very different designs of locomotive. Out along the Great Western's main line from Paddington to Bristol and Exeter, William Dean's famous single-driver type was preferred, with the recently developed 4-4-0 designs covering the more arduous later stages over the notorious South Devon banks. While this arrangement was suitable for the traffic needs of the day, future development of the company and its locomotive designs demanded long-term planning. Churchward's vision of the GWR's motive power needs were based on long-term forecasts in traffic growth, taking in the developments in motive power design overseas, particularly the USA, and were very much more radical than traditional thinking.

William Dean had already left his mark in the history books with some classic designs, of which his single-driver type of the 1890s was among the most successful and attractive of the designs that were fashionable at the time. These locomotives entered service at the end of the company's broad gauge era, and were originally of 'convertible' design. Only three years later the first of a series of 4-4-0 types appeared, intended for working

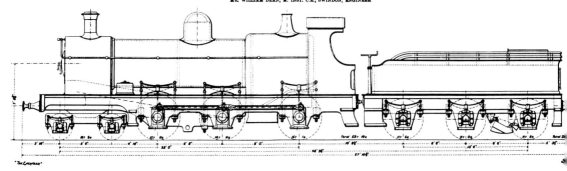

TEN-WHEELED MINERAL ENGINE, GREAT WESTERN RAILWAY

MR. WILLIAM DEAN, M. INST. C.E., SWINDON, ENGINEER

over the hilly routes west of Exeter. In their turn these first 4-4-0s, the Duke or Devon class, were followed before the turn of the century by Bulldog, Badminton, and Atbara classes, with varying degrees of Churchward's influence on their design and construction. On the freight side, Dean had produced the 260 Class 2301 0-6-0s – the Dean Goods – which were still being built in the 1890s, and for which he may perhaps be remembered most. While his goods engines

The classic Dean single, as exemplified by No 3047 *Lorna Doone*, with polished brass dome, gleaming copper capped chimney, and graceful lines, soon to disappear under Churchward's austere but efficient practices. *Locomotive & General Railway Photographs*

Two-cylinder mixed traffic 4-4-0s for South Devon. As seen here on No 3252 *Duke of Cornwall*, the Dukes first took to the rails with wooden bogie and tender wheel centres. *Locomotive & General Railway Photographs*

No 3312 *Bulldog* was the first to appear with some indication of Churchward's influence – along with a Standard No 2 parallel boiler came a Belpaire firebox, but it still sported the huge brass dome. *Locomotive & General Railway Photographs*

were conventional by and large, in 1897 Dean, despite an awareness that his health was deteriorating, presided over the introduction over the first GWR 4-6-0, No 36. This particular design, an expansion of the 4-4-0 type, was one of the most curious, ungainly looking locomotives to emerge from Swindon, and not a fitting representative of the elegant Victoriana so characteristic of the Dean era.

'The Crocodile,' alias No 36, was the first two-cylinder 4-6-0 on the GWR, in slightly modified form here, as the boiler feed clacks have been removed from their original position. This locomotive was built under Dean's direction in 1896.

The First Great Western 4-6-0

This first GWR 4-6-0 was a goods engine, and for some time remained a solitary example of the type, until the Krugers appeared in 1899. By the standards of 1896, No 36 was hailed as massive; double frames, large round-top firebox, and substantially larger than other goods types, in every aspect. It was intended that the use of No 36 would reduce the need for double and even triple heading through the Severn Tunnel. Built in August 1896, it was withdrawn in December 1905, and there is a theory that, under its nickname, 'The Crocodile', this locomotive may not have travelled far from Swindon, although this idea may not be supported in fact. Its tender survived for many years, and was still in use around the time of the second world war.

The design of No 36 was entirely under Dean's authority, and included the largest boiler ever built to his ideas. The 14ft 0in barrel was 4ft 6in in diameter, and constructed in two rings, within which were housed 150 of the 2½in diameter corrugated Serve tubes. According to one contemporary press report, Mr Dean spoke very highly of the Serve tubes, which apparently gave no trouble, in a highly efficient boiler. To carry the brick arch across the 5ft 0in wide firebox, water tubes were installed. The greater than normal width of the firebox (5ft 10in over the outer wrapper plates) was possible because the inner frames of the double mainframes stopped at the front end of the firebox. The firegrate, 30.5sq ft in area, was unusual in having a flat surface at the back and front, and a steeply sloping centre section connecting the two. This arrangement was possible only because of the 4ft 6in coupled wheels. At the opposite end of the boiler, a fully extended smokebox was fitted, waisted-in to fit between the frames, below which were the two 20in × 24in cylinders, actuated by Stephenson valve gear.

The coupled wheels, whose size was officially quoted at 4ft 6in diameter, were increased to 4ft 7½in by using thicker tyres. And while the locomotive was fitted with the relatively new Dean bogie, the 2ft 8in bogie wheels had wooden centres, like the Mansell coach wheels, and the Duke class 4-4-0s. It was a distinctly odd-looking 4-6-0, not at all like the modern 4-6-0s introduced only a few years later, but for all that No 36 was quite a powerful machine, exerting a tractive effort of 24,933 lb, with a boiler working pressure of only 165 lb/sq in.

By the end of the 19th century, the GWR had acquired some excellent passenger locomotives to meet its traffic demands, and at least one prototype heavy goods 4-6-0. The need for a truly mixed-traffic type was not so apparent in the 1890s, but for the company's future substantial expansion of traffic it was Churchward and not Dean who was to provide these much needed designs. The way forward for the locomotive department was already becoming obvious, notably in the many changes taking place on the numerous 4-4-0 classes. The next generation of both mixed traffic and passenger locomotives, and many others after that, owed their existence to the GWR's next prototype 4-6-0, No 100, forerunner of the illustrious Saint class.

4-6-0 No 36 – leading details

Running number	36		
Lot number	106		
Works No	1551		
Built	8/1896		
Withdrawn	12/1905		

Wheel diameter			
– coupled	4ft 6in/4ft 7½in		
– bogie	2ft 8in		
Wheelbase (locomotive only)	25ft 0in		

Axle load	Tons Cwt leading	Tons Cwt driving	Tons Cwt trailing
– coupled	16 11	15 12	15 01
	Tons Cwt		
– bogie	12 06		

Cylinders		
– number	2	
– size	20in × 24in	

Boiler		
– length	14ft 0in	
– diameter (o/s)	4ft 6in/4ft 7in	

Tubes		
– number	150	
– size	2½in	
(Serve – internally ribbed)		

Working pressure	165lb/sq in

Firebox – length	7ft 0in

Grate area	30.5 sq ft

Heating surface	
– tubes	1402.06 sq ft
– firebox	115.83 sq ft
– total	1517.89 sq ft

Tractive effort	24,933 lb

Fuel capacities	
– coal	5 tons
– water	2,600 gallons

Weights	
– locomotive	59 tons 10 cwt
– tender	35 tons 15 cwt
– total	95 tons 05 cwt

CHAPTER TWO

CHURCHWARD'S TWO-CYLINDER 4-6-0s

By the time that George Jackson Church-ward was finally charged with overall responsibility for the Great Western locomotives much experimental work had already been done. His plan for a range of standard locomotive designs had been formulated the year before he took office, and in Dean's last days the company's first two-cylinder 4-6-0 appeared in the shape of the prototype, No 100.

It has been suggested that Churchward's work was much influenced by contemporary American practice, and a comparison of the relative simplicity of English and American designs of the 1890s might serve to illustrate this idea. Churchward's standard designs were far simpler in construction and layout than almost anything that had gone before, and represented a successful combination of the latest thinking. British engineers it seems had persisted in the extension, not development, of cylinder and valve gear layout, frequently within double frames, and in form similar to that adopted by Stephenson himself many years earlier.

Un-named prototype No 100, with original parallel boiler, in operation early in 1902. *Historical Model Railway Society*

While such arrangements were acceptable, and often necessary for the earlier four-wheeled engines, it had been extended to include six- and eight-wheelers. Double frames and cylinders inside those frames were an impossible concept for Churchward, and would have imposed considerable restraints on design, in construction, and in service. Bearing in mind the anticipated and actual growth in traffic on the GWR, Churchward was, unlike many other railway companies' engineers, designing motive power for future and not just current needs.

In 1901, Churchward's approach to the design of his new range of standard locomotives was in many ways paralleled almost half a century later by the newly formed British Railways organisation. In this, the GWR chief mechanical engineer was looking around to see what others were doing, adapting and tailoring new ideas to suit the needs of the GWR. While the simple layout and construction techniques were an obvious American influence, European features, in particular the success of the French railways' compound designs was another. Compounding would have introduced complications, so Churchward proceeded to design simple expansion engines, with

improved mechanical design, looking to improve their thermal efficiency still further at a later date.

Other details standardised by Church-ward included the use of Stephenson valve gear with outside cylinders and inside frames only, and the definition of only three sizes of coupled wheel: 6ft 8½in, 5ft 8in, and 4ft 7½in. Churchward's most memorable contribution to locomotive design will always be in the development of the boiler. The principles embodied in GWR boiler design were outlined in a paper he presented to the Institution of Mechanical Engineers in March 1906. By that time however, some very successful and far-reaching develop-ments had already taken place.

The First Prototype – No 100

In February 1902 No 100 a two-cylinder 4-6-0 emerged from Swindon Works, under lot No 132. It was a curious looking machine, dramatically different from all that had gone before under William Dean's leadership. Gone were the double frames, the flowing lines of the singles and early 4-4-0s, the shin-ing brass dome, and all the ostentation of the Victorian era. In its place, and in an austere way No 100, the first express passenger 4-6-0 on the GWR, brought the railways smartly into the 20th Century. It was by comparison with the many earlier designs a stark looking locomotive, including (oddly perhaps) a parallel boiler with a working pressure of only 200 lb/sq in.

With its saturated boiler, 14ft 8in long overall, No 100 was fitted with a steel fire-

No 100 with its later name *William Dean*, now carrying a short taper boiler, and nameplates on the leading splashers. Sheet metal fenders have appeared on the tender, replacing the earlier coal rails. *Locomotive & General Railway Photographs*

box, having a Belpaire casing, raised above the adjoining boiler barrel, and was an ex-perimental design. From June to November 1902, the locomotive was named simply *Dean*. In June 1903, just over a year after its completion, No 100 was carrying the name *William Dean*, and a short cone (the rear tapered boiler ring) new Standard No 1 boiler. For the new passenger locomotives, slide valves paired with the 18in × 30in cylinders were out of the question, and Churchward initially specified 6½in diam-eter, double-port piston valves. The reason for the use of this type and size of piston valve was also essentially experimental, and following investigations into the early per-formances of No 100, the valves were opened out to 7in, and later 7½in diameter, in an effort to reduce axlebox knock. Follow-ing further investigations, it was decided to standardise on 10in diameter, single-port piston valves. At the time, there had been little experience of large piston valves in this country, and Churchward's approach re-sulted ultimately in the simple design prov-ing more efficient overall for the company's purposes. The final style of 7½in diameter double-port valves were retained by No 100, and with detail differences in framing, valve gear and cylinders, the locomotive was a distinctive member of the later Saint class.

Other detail alterations to No 100 included the fitting of a superheater in April 1910, top feed in November 1913, and a four-cone ejector in May 1915. The safety valve bonnet, sitting on the long cone fully tapered boiler fitted from 1912, was changed to a shorter design from May 1927. Although a 4,000-gallon tender was officially paired with this locomotive, changes resulted in the more common pairing with 3,500-gallon tenders.

The first prototype 4–6–0
No 100 – leading details

Running No	100 (later 2900)
Swindon Works Lot No	132
Works No	1928
Built	2/1902
Withdrawn	6/1932

Wheel diameter (coupled)	6ft 8½in
Wheel diameter (bogie)	3ft 2in
Wheelbase (locomotive only)	27ft 2in

	Tons	Cwt
Axle load – coupled, – leading	17	10
– driving	18	00
– trailing	17	00
Axle load – bogie	15	06
Cylinders – number	2	
– size	18in × 30in	

Boiler – length	14ft 8in
– diameter (o/s)	4ft 10¾in & 5ft 0in
– tubes – number	287
– size	2in (o/d)
Working pressure	200lb/sq in

| Firebox – length | 9ft 0in |
| – width | 5ft 6⅛in |

| Grate area | 27.62sq ft |

Heating surface – tubes	2252.37 sq ft
– firebox	157.94 sq ft
– Total	2410.31 sq ft

| Tractive effort | 20,530lb |

| Fuel capacities – coal | 5 tons |
| – water | 4,000 gallons |

Weights (in working order)	Tons	Cwt
– locomotive	67	16
– tender	43	03
– Total	110	19

Visually, No 100 contrasted dramatically with even some of the more austere Churchward modified designs of William Dean. The new locomotive's running boards, attached in the conventional manner to the frames, dropped abruptly at the leading end to the front footplate, and the cab was reached by way of three large footsteps. The American influence obviously did not extend to the cab, which was hardly worth the name, but certainly typical of Churchward thinking. The abrupt steps in footplate levels, and a lack of curvature at the cab end contributed to this and the next prototype 4-6-0, No 98's, rather unattractive appearance. However, unlike No 100 the next two-cylinder 4-6-0 from Swindon was much more characteristic of Churchward, and the first of the new standard locomotives of the Great Western Railway.

No 98 – a true Churchward design

The second prototype two-cylinder 4-6-0 to emerge from Swindon, in March 1903, was the first true Churchward design. In layout, No 98 followed some of the basic principles incorporated in its predecessor, still only a year old, but there were some very important and major changes. This new locomotive encapsulated many of Churchward's design principles, which with some development in detail remained characteristic of GWR locomotive practice almost until nationalisation. No 98 was built at Swindon to Lot No 138, and carried works number 1990; like *William Dean*, No 98 was first set to work without a name.

Among the most obvious changes in the appearance of this forerunner of the Saints was the fitting of the No 1 Standard boiler from new, with initially a short tapered cone and trapezoidal Belpaire firebox. As with the earlier prototype, No 98 was originally paired with a 4,000-gallon tender. The construction of the footplating with its abrupt changes of level was the same, including the diminutive looking cab, whose design really appears as though it were an afterthought. No 98 was designed at the same time as the new standard heavy freight locomotive, which subsequently became the 28XX 2-8-0 type, with which some difficulties encountered resulted in minor dimensional changes at the front end of No 98.

The second of the prototype 4-6-0s, No 98 in its original form, with short cone Standard No 1 boiler, waits at Bath. This locomotive later became the last numbered member of the Saint class, No 2998. *Locomotive & General Railway Photographs*

The second prototype 4-6-0, No 98 – leading details

Running No	98 (Later 2998)
Lot No	138
Works No	1990
Built	3/1903
Withdrawn	6/1933
Wheel diameter (coupled)	6ft 8½in
Wheel diameter (bogie)	3ft 2in
Wheelbase (locomotive only)	27ft 1in

	Tons	Cwt
Axle load – coupled – leading	16	08
– driving	18	00
– trailing	17	10
Axle load (bogie)	16	08
Cylinders – number	2	
– size	18in × 30in	

Boiler – Length	14ft 10in
diameter (o/s)	4ft 10¾ tapering to 5ft 6in
Tubes – number	250
– size	2in (o/d)
Working pressure	200lb/sq in
Firebox – length	9ft 0in
– width	5ft 9in
Grate area	27.22 sq ft
Heating surface – tubes	1988.65 sq ft
– firebox	154.39 sq ft
– total	2143.04 sq ft
Tractive effort	20,530lb
Fuel capacities – coal	5 tons
– water	4,000 gallons

Weights	Full		Empty	
	Tons	Cwt	Tons	Cwt
– locomotive	68	06	63	03
– tender	43	03	19	05
– Total	111	09	82	08

The boiler fitted to No 98 was built in two rings, the rearmost ring being coned outside; this taper was visible from the front ring (parallel) to the firebox. Later modifications included a more conventional tapered boiler, extending from smokebox to firebox. The majority of what became the Saint class locomotives were fitted with this latter type of boiler, in their original form. No 98 carried its saturated, short taper boiler until 1908, when it was reboilered and fitted with a long cone boiler, in which guise it was superheated in September 1911. The Belpaire firebox fitted to this locomotive was trapezoidal in plan, and shaped to fit between the frames, tapering from a maximum outside width of 5ft 9in, to a minimum of 4ft 0in at the cab end. On the GWR, a Churchward-style firebox also tapered from a maximum height at its junction with the boiler, sloping down towards the cab spectacle plate.

Perhaps the most important changes were made at the front end of the locomotive, where the two 18in diameter cylinders with their 10in diameter single-port piston valves above were incorporated in two castings, which also included half-saddles to carry the smokebox. Churchward's cylinder design, in which the steam chest partially surrounded the cylinders, keeping them as warm as possible, was originally designed for use with saturated boilers. This form of construction also prevented the use of full length steel plate frames, so in Churchward practice the mainframes stopped just behind the cylinders and bar frame extensions were bolted on to run under the cylinder casting assembly to the front buffer beam. The frame structure in Churchward's prototype 4-6-0 looks weak in this area, but the arrangement was standardised for construction until the arrival of F. W. Hawksworth as chief mechanical engineer.

The valve gear adopted as standard for the 2-cylinder locomotives was the Stephenson type, and although the lengths of the rods varied according to the wheel spacing of the various designs, the eccentrics and expansion links were standard components. This applied to No 98 and the prototype locomotives recently built or under construction at that time. These standard components remained unchanged until the end of steam traction in this country! There was something of a problem in the connection of the valve rods for the outside cylinders and the inside motion. In contemporary American practice it was simple enough to arrange for the inclusion of rocking shafts passing through large holes in the bar frames and connecting directly with the valve rods. In Churchward's design this was not possible and the necessary rocking shafts were carried above the top edge of the plate frames to avoid any weakening of the mainframe structure.

While there were a number of features of contemporary American design in No 98, this heavy influence was not so apparent in the locomotive's outward appearance. Coming out in March 1903, No 98 ran without a name for four years until March 1907, and was the most powerful passenger locomotive that the GWR possessed for a time. When first built it was rumoured that it was to be named *Persimmon*, but the name finally bestowed was *Vanguard*, which was only carried until December 1907 when it was changed to *Ernest Cunard*. By the time No 98 received a name, the original boiler pressure of 200 lb/sq in had been increased to 225 lb/sq in and later modifications applied in common with other members of the eventual Saint class.

The Atlantic Era

Saint class Nos 2971–2990 were the subject of major alterations. A number were built as 4-4-2s, to enable a more direct comparison to be made with the French compounds then on trial. The number series of these locomotives was originally 171–190, with No 171 *Albion* built in December 1903 as 4-6-0 with the standard 225 lb/sq in boiler pressure. *Albion* also had some minor alterations to the firebox, resulting in a reduction of the heating surface to 154.26sq ft, while the increased boiler pressure improved the figure for tractive effort, to 23,090 lb. This locomotive, which in effect was third in the Saint series, ran in this form until its conversion to a 4-4-2 in October 1904.

The conversion consisted basically of the provision of a trailing axle under the firebox, with 4ft 1½in diameter wheels, and outside bearing axleboxes. Spaced at a distance of 8ft 3in from the trailing coupled axle, this extended the rigid wheelbase by 6in and caused some minor changes in the axle loadings. Extra plates had been fitted to *Albion* during construction to facilitate alteration to the alternative wheel arrangement, and this feature was incorporated in a further thirteen 4-4-2s, built in this form in 1905. Nineteen locomotives were built at Swindon, to Lot Nos 154 and 158, being numbered 172 to 190. Nos 172, 179 and 190 came out as Atlantics, Nos 173 to 178 as 4-6-0s. The locomotives were all built with Standard No 1 boilers with short tapered cone; seven were superheated while still running as 4-4-2s. Changes resulted in the fitting of boilers with long tapered cones to all but three locomotives in the Atlantic form.

The trials with the imported French compounds 4-4-2s Nos 102 to 104 proved that for the GWR at least little would be gained by the adoption of compounding, especially with regard to fuel consumption. However it did emerge that the use of divided drive in four-cylinder locomotives would tend to result in longer intervals between shopping, and this layout would be borne in mind for future multi-cylinder designs. The bogie of the French compounds appeared to be a contributing factor in its smooth running, and a derivative of this design was fitted to Atlantic No 184. By the end of 1905 the compound system was proving to offer little advantage, and certainly the two-cylinder

4-6-0s with their long-travel valves were equally as economic in service as the complex compounds. The 4-6-0s also demonstrated an advantage in adhesion, which was an important consideration, particularly where services on the hillier routes, such as those of South Devon, were to be worked.

Of the Atlantics, No 187 had a curved fall plate at the leading end, introduced perhaps to counter the critics of Churchward's severe designs. In any event, when the locomotives were converted to 4-6-0s between April 1912 and January 1913 the mainframes had been altered to allow these curved connections between the running boards and footplate at

For comparative trials with the imported French Atlantics, one batch of two-cylinder locomotives was built with a 4-4-2 wheel arrangement. No 188 *Rob Roy* is seen here in its original form. *Historical Model Railway Society*

front and cab ends. The six 4-6-0s outshopped in 1905 with the Atlantics were slightly heavier, and were paired with smaller, 3,500-gallon tenders. The Atlantic series carried names taken from the novels of Sir Walter Scott, and while collectively they were all included like the prototype loco-

One of the Lady series, No 2906 *Lady of Lynn*, with the earlier form of tall safety valve covers, abrupt change of level at the smokebox end of the running plate, and large cab footsteps. Initially, this locomotive was fitted with a short taper boiler. *Lens of Sutton*

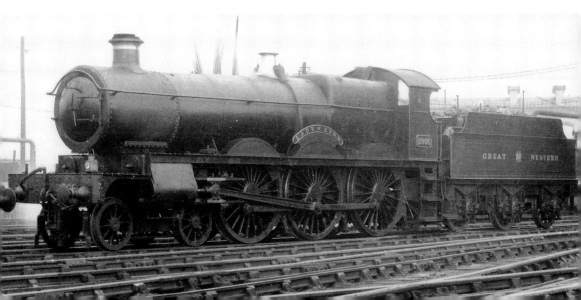

motives in the Saint class, they were also known as the Scotts. Three more series can be identified under the collective description of Saint class – the Ladies, the Saints *per se*, and Courts.

The Ladies

The next batch of GWR passenger locomotives to take to the rails was the Lady series of ten, built in May 1906 to Swindon Lot No 164. These too had the severe lines of the earlier Churchward 4-6-0s and 4-4-2s, but introduced some distinctive new features. They carried running numbers 2901 to 2910, and for their earliest allocations, six went to Paddington, two to Bristol, and one each to Newton Abbot and Plymouth.

Boilers for the Lady series were built under different orders, and resulted in short taper Standard No 1 boilers on Nos 2904/5/6, with short smokeboxes; long taper boilers with medium length smokeboxes appeared on Nos 2902/3/7–10. No 2901 was built with a short taper standard No 1 boiler, but achieved the further distinction of being the first British steam locomotive to be fitted with the Schmidt pattern firetube superheater. This locomotive had had its boiler pressure reduced to 200 lb/sq in, and the cylinder diameter increased slightly to 18⅜in, and there was some reduction in tractive effort. The altered boiler details for No 2901 were as follows:

Boiler – type	Standard No 1 (short cone)
– working pressure	200 lb/sq in
Tubes – small	130 × 2in o/d
– large	24 × 4¾in o/d
Superheater elements	24 (three rows of 8 elements)
Heating surface – boiler	1485.96sq ft
– firebox	154.94sq ft
– total evaporative	1640.90sq ft
– superheater	307.52sq ft

From September 1909, this locomotive carried a long taper Standard No 1 boiler without superheater and finally from May 1912 a standard Swindon superheater boiler. Others in the Lady series were given 18½in diameter cylinders, both short and long taper boiler versions, resulting in an increase in their tractive effort to 23,382 lb.

The Saints

In 1907, the first locomotives to be given 'Saint' nameplates emerged from Swindon, with those built to Lot No 170 receiving the running numbers 2911 to 2930. They were also the last Swindon builds to carry works plates on the locomotive. The boiler carried by the Saints was a long taper, Standard No 1, but they were still unsuperheated and it was not until the end of the decade that the Swindon design standard superheaters were

A true Saint, No 2927 *Saint Patrick*. Although the tall safety valves remain, a much smoother overall appearance is given with the abandonment of the abrupt change of levels for the running plates. *Lens of Sutton*

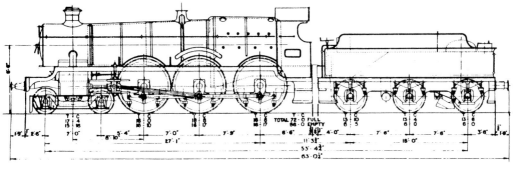

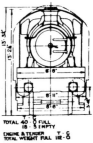

Saint class 4-6-0

fitted on a large scale to the two-cylinder 4-6-0s.

Among the most important changes introduced with the 2911–2930 series of locomotives, was the adoption of screw reverse. Churchward had continued to use lever reverse on the large 4-6-0s for some time, on the early prototypes, and with the Scott and Lady series. The difficulty in making small adjustments in cut-off with large hand levers led to the development of the practice of driving on the regulator, and making cut-off adjustments at low speed. This almost fixed cut-off did not prevent the locomotives concerned from doing excellent work, but the eventual use of screw reverse introduced more flexibility, and gave the footplatemen a finer degree of control over the locomotives.

In 1908 No 2922 had the Swindon No 2 superheater fitted experimentally, with 4¾in diameter flues, 108 elements, and a heating surface of 275sq ft. Two years later the Standard Swindon No 3 superheater was introduced, first fitted to Nos 2913 and 2920, but eventually to all, including the earliest 4-6-0s and 4-4-2s, by November 1912. Other detail alterations carried out during World War I included an experimental rocking grate on No 2914, and alterations to the top feed mounting to No 2917.

The Ladies and the Saints were by 1910 establishing a well deserved reputation as main line passenger engines, capable of heavy haulage and high speeds. They were coming to the end of Churchward's experimental period and were followed in 1911 by the final development of the two-cylinder 4-6-0 under his leadership, the Court series.

The Courts

The 25 locomotives built to this design came out to Swindon Lot Nos 185, 189, and 192, between 1911 and 1913. They were superheated from new, with the now standard top feed to the long taper No 1 boilers. The Court series locomotives were all built with long smokeboxes. The cylinders were carried in the horizontal position, on the same centre line as the axles of the coupled wheels. There were some minor variations in the layout of the boilers, and from No 2941 all members of the class were fitted with 18½in diameter cylinders. This subsequently became the standard for all the Saint class 4-6-0s, with tractive effort up to 24,395 lb. Some boiler changes resulted in Court series running with short taper No 1 boilers, demonstrating the interchangeability of Churchward's boiler designs.

The Court series engines carried the names of well-known houses within GWR territory, some being the homes of the com-

The final version of the Saints, the Court series. No 2930 *Cefntilla Court* with the fully tapered Standard No 1 boiler. *Lens of Sutton*

pany's directors. Five engines built to lot 192 in 1913, and numbered 2951–2955, were originally part of a batch of 10 to be built under that lot number, although the remainder, to be numbered 2956–2960, were cancelled in November 1912. Construction of this final series of Saint class engines took place at the same time as a renumbering scheme was being introduced. The main purpose of the scheme, which took effect from December 1912, was to tidy up the allocation of running numbers generally. The system adopted was also intended to convey an indication of the wheel arrangement and number of cylinders – for example X0XX series were the four-cylinder 4-6-0s (Kings, Castles, Stars), X9XX two-cylinder 4-6-0s (Saints, Halls), X8XX two-cylinder mixed traffic 4-6-0s (Granges, Manors). The Hawksworth two-cylinder County class 4-6-0s broke this sequence when they were given numbers in the 10XX series. All the prototype locomotives were allocated running numbers in accordance with this scheme, Nos 98, 100, and 171 to 190 becoming 2998, 2900, and 2971 to 2990 respectively.

Operations and later development

No fewer than 13 diagrams were issued to cover the Saints as 4-6-0s, and two more for the 4-4-2 versions. The locomotives were power class C and restricted to red routes. By 1913, the 92 Saints were allocated as follows: Old Oak Common (London) 14, Bristol 18, Plymouth 4, Newton Abbot 7, Exeter 5, Wolverhampton 11, Gloucester 5, Cardiff 9, Fishguard 4.

They were used on main line expresses, especially with the early allocations to London and in Devon, although they were seen even earlier in the Wolverhampton and west Wales areas. One of the Lady series, No 2902 *Lady of the Lake* worked the first GWR two-hour train from Birmingham to London, on 1 July 1910. Just before the train's departure, a well-wisher threw a horseshoe onto the footplate. This was later mounted, and was carried on the locomotive until its withdrawal in 1949.

Saint class 4-6-0s demonstrated from new their ability to show a clean pair of heels, with No 2903 reportedly reaching 120mph, unofficially, on a trial trip, before the locomotive was properly run-in. In later years No 2903 took a test train from Paddington to Plymouth via Westbury in 3¾ hours, with a similar run to Exeter in less than three hours. By the 1920s some reorganisation in South Wales resulted in a large batch of Saints being moved to Llandore, although Old Oak Common and Bristol still retained the largest allocations.

The 1920s also saw some interesting changes in detail on this class – the introduction of shorter safety valve bonnets from

1927, and whistle shields from around 1925. No 2933 had the blastpipe position altered and the chimney moved forward, in which guise like the Star class 4-6-0 No 4060 it ran from 1923 to 1925. Another of the Saint series, No 2930, was fitted with adjustable hornblocks, while No 2947 *Madresfield Court* was provided with cylinder by-pass valves and experimental valve spindle guides. Beginning with No 2903 in November 1930, the fitting of new front ends – complete cylinder assemblies and frame extensions – resulted in the appearance of outside steampipes. Many of the Saint class were thus altered between 1930 and June 1948, but 36 locomotives went to the scrapyard in basically their original form.

A really unusual modification was the rebuilding in 1931 of No 2935 *Caynham Court* with a new set of cylinders and rotary cam valve gear. No 2935 was the only GWR locomotive to have poppet valve motion, and one of the few departures from the simple standardised designs laid down by Churchward. It goes without saying that this locomotive was modified after Churchward's retirement, when there had been some loosening of his fairly rigid policies. The unusual arrangement adopted on No 2935 did not prove to have any significant advantages, much like the earlier comparisons with the

French 4-cylinder compounds, and it was not repeated.

The fitting of new front ends to the Lady series considerably reduced the number of this type with the sudden changes in levels of the running boards. Those which had been built as Atlantics and later rebuilt as 4-6-0s were also modified to have a curved connection between front footplate and running boards adjacent to the smokebox, leaving only Nos 2900/1/4/7/9/10/71/3–8/98 in their original form. There were also some minor differences between the various series which went to make up the Saint class as a whole, including chimneys, modifications to nameplate fittings, and lamp irons carried on smokebox doors. The GWR automatic warning system gear was fitted early in the life of the Saints, Nos 2901/8/86 receiving the equipment in 1908. Speedometers began to appear from about 1937, although unlike the AWS it was not fitted to the entire class; at least 34 locomotives were equipped. Since the boilers were interchangeable with the newer Grange class 4-6-0s after that type began to appear, some Saints could be found

No 2931 *Arlington Court*, one of the last series of Saint class locomotives, and the final development of the two-cylinder 4-6-0 under Churchward, is here seen under test at Swindon in June 1935. *L. Hanson*

Saint Class No 2943 *Hampton Court* (screw reverse) and No 2979 *Quentin Durward* (lever reverse) awaiting scrapping at Swindon in 1951. *A. R. Brown*

with Grange type chimneys as well, from about 1939.

While Castle class 4-6-0s had displaced the Saints on the heaviest workings, during the 1920s and 1930s the latter continued to be employed on the GWR's principal main line services. During the second world war they earned the nickname Hereford Castles, sharing 16- and 17-coach trains with members of the Castle class, working between Shrewsbury and Pontypool Road.

The unofficial record run of No 2903 and its allocation to the first of the two-hour Birmingham trains in 1910, was joined by another 'first' in 1923. On this occasion, the GWR's new prestige train, the Cheltenham Flyer, was taken on its first 75-minute schedule (Swindon to Paddington) by No 2915 *Saint Bartholomew*.

Withdrawal of the Saints began in August 1931, when No 2985 was taken out of service, joined before the end of the decade by another 21 members of this illustrious class. The two prototype locomotives, Nos 2900 and 2998 were also taken out of stock in June 1932 and June 1933 respectively. Further withdrawals took place in the 1940s, with

another seven going before Nationalisation in 1948, when 47 were taken-over by British Railways.

The introduction of Hawksworth's County class 4-6-0s in 1945 eventually led to the further demise of the Saints, with their relegation to stopping passenger and main line relief trains. Under GWR ownership in 1947, Saints were allocated to Swindon, the West Country and South Wales, although the largest allocation was at Hereford, where eight remained, with significant numbers at other North and West Midlands sheds. One of the Hereford-based locomotives on withdrawal in 1952 had completed more than two million miles. The actual figure of 2,076,297 miles, was the highest recorded for any Great Western locomotive, and represented yet another 'first' achieved by this class.

The most important modification to any of the Saint class 4-6-0s was the reconstruction of No 2925 *Saint Martin* in 1924 with 6ft 0in coupled wheels, and other structural changes including the fitting of the Collett side window cab. This prototype for the highly successful Hall class was produced under the direction of C. B. Collett, and although a similar design had been outlined many years earlier by Churchward, the rebuild was carried out essentially to satisfy the demands of the running department for a locomotive of this capability. The prototype Hall, eventually renumbered 4900 is described in more detail in Chapter 4. The Saint class diagrams were:

Diagram	Wheel arrangement	Details
B	4-4-2	Atlantics as built
F	4-4-2	Atlantics, superheated
C	4-6-0	No 100 as fitted with taper boiler, 1903
D	4-6-0	No 98 as built
E	4-6-0	No 171 as built
F	4-6-0	Nos 2902-2910 as built with fully tapered boiler
G	4-6-0	No 2901 fitted with Schmidt type superheater
I	4-6-0	Nos 2911-2930, Saint series, as built
L	4-6-0	No 2922 with first Swindon No 2 superheater
O	4-6-0	No 2913, with Swindon No 3 superheater
P	4-6-0	No 100, with Swindon No 3 superheater
R	4-6-0	Nos 2931-2940, Court series, as built
V	4-6-0	29XX Saint class, with standard 18½in cylinder
X	4-6-0	No 2925 as rebuilt to prototype Hall
A4	4-6-0	No 2935 as rebuilt with rotary cam poppet valve gear

Forty-seven locomotives passed into BR ownership, although neither of the two prototype 4-6-0s Nos 100 and 98 survived.

Six originally built as Atlantics were among them. Six were withdrawn fairly soon after nationalisation in 1948. Of those six, five had not been fitted with new cylinders, front ends, and outside steam pipes, although No 2937 *Clevedon Court* was the only Saint to be altered in BR days, emerging from Swindon with outside steam pipes in June 1948. The solitary Caprotti-fitted Saint, No 2935 *Caynham Court* was still in service, disappearing under the breaker's torch in December 1948.

In BR ownership the Saints were power class 4P, and were turned out in the standard mixed traffic lined black livery, although not all received the new colour scheme before being withdrawn. Smokebox numberplates were applied to Nos 2906/ 8/ 12/ 15/ 20/ 6/ 7/ 31–34/ 36–38/ 40/ 42–45/ 47–54/ 79/ 81/ 87. Earlier Nos 2903/29/51/55 had carried the temporary W suffix to the running number. By 1950 only 29 remained in service, at the following locations: Bristol 4, Swindon 7, Weymouth 1, Banbury 1, Leamington 1, Tyseley 1, Chester 3, Gloucester 2, Hereford 3, Newport 3, Cardiff 3.

Completing its days at Hereford in 1952 No 2920 *Saint David*, built in September 1907, recorded no fewer than 2,076,297 miles. Although No 4037 – a Star class locomotive rebuilt as a Castle – exceeded that figure, No 2920 was basically as built throughout its life.

Saint class 4-6-0 building and withdrawal dates

No	Name	Built	First depot*	Withdrawn	Last depot
2900	William Dean	2/02	Bristol	6/32	Chester
2901	Lady Superior	5/06	Bristol	4/33	Chester
2902	Lady of the Lake	5/06	Paddington	8/49	Leamington
2903	Lady of Lyons	5/06	Plymouth	11/49	Tyseley
2904	Lady Godiva	5/06	Paddington	10/32	Shrewsbury
2905	Lady Macbeth	5/06	Newton Abbot	4/48	Cardiff
2906	Lady of Lynn	5/06	Paddington	8/52	Cardiff
2907	Lady Disdain	5/06	Bristol	7/33	Newport
2908	Lady of Quality	5/06	Paddington	12/50	Swindon
2909	Lady of Provence	5/06	Paddington	11/31	Stafford Road
2910	Lady of Shalott	5/06	Paddington	10/31	Westbury
2911	Saint Agatha	8/07	Bristol	3/35	Taunton
2912	Saint Ambrose	8/07	Paddington	2/51	Weymouth
2913	Saint Andrew	8/07	Paddington	5/48	Swindon
2914	Saint Augustine	8/07	Paddington	1/46	Weymouth
2915	Saint Bartholomew	8/07	Paddington	10/50	Chester
2916	Saint Benedict	8/07	Bristol	7/48	Tyseley
2917	Saint Bernard	8/07	Paddington	10/34	Shrewsbury
2918	Saint Catherine	8/07	Cardiff	2/35	Leamington

No	Name	Built	First depot*	Withdrawn	Last depot
2919	Saint Cuthbert	9/07	Cardiff	2/32	Taunton
2920	Saint David	9/07	Cardiff	10/53	Hereford
2921	Saint Dunstan	9/07	Bristol	12/46	Banbury
2922	Saint Gabriel	9/07	Plymouth	1/45	Severn Tunnel Junction
2923	Saint George	9/07	Plymouth	10/34	Bristol, Bath Road
2924	Saint Helena	9/07	Paddington	3/50	Hereford
2925	Saint Martin	9/07	Newton Abbot	12/24†	Paddington
2926	Saint Nicholas	9/07	Paddington	9/51	Chester
2927	Saint Patrick	9/07	Cardiff	12/51	Swindon
2928	Saint Sebastian	9/07	Exeter	8/48	Westbury
2929	Saint Stephen	9/07	Cardiff	12/49	Bristol
2930	Saint Vincent	9/07	Bristol	11/49	Chester
2931	Arlington Court	10/11	Exeter	2/51	Bristol
2932	Ashton Court	10/11	Newton Abbot	6/51	Tyseley
2933	Bibury Court	11/11	Exeter	1/53	Leamington
2934	Butleigh Court	11/11	Newton Abbot	6/52	Swindon
2935	Caynham Court	11/11	Laira	12/48	Swindon
2936	Cefntilla Court	11/11	Cardiff	4/51	Newport
2937	Clevedon Court	12/11	Fishguard	6/53	Hereford
2938	Corsham Court	12/11	Fishguard	8/52	Gloucester
2939	Croome Court	12/11	Exeter	12/50	Bristol, Bath Road
2940	Dorney Court	12/11	Fishguard	1/52	Cardiff
2941	Easton Court	5/12	Paddington	12/49	Westbury
2942	Fawley Court	5/12	Paddington	12/49	Bristol, Bath Road
2943	Hampton Court	5/12	Laira	1/51	Cardiff
2944	Highnam Court	5/12	Paddington	11/51	Hereford
2945	Hillingdon Court	6/12	Bristol	6/53	Cardiff
2946	Langford Court	6/12	Bristol	11/49	Westbury
2947	Madresfield Court	6/12	Paddington	4/51	Swindon
2948	Stackpole Court	6/12	Cardiff	11/51	Bristol, Bath Road
2949	Stanford Court	5/12	Wolverhampton	1/52	Swindon
2950	Taplow Court	5/12	Paddington	9/52	Bristol, Bath Road,
2951	Tawstock Court	3/13	Cardiff	6/52	Gloucester
2952	Twineham Court	3/13	Cardiff	9/51	Severn Tunnel Junction
2953	Titley Court	3/13	Gloucester	2/52	Chester
2954	Tockenham Court	3/13	Newton Abbot	7/52	Swindon
2955	Tortworth Court	4/13	Newton Abbot	3/50	Weymouth
2971	Albion	12/03	Paddington	2/46	Swindon
2972	The Abbot	2/05	Plymouth	3/35	Cardiff
2973	Robins Bolitho	3/05	Paddington	7/33	Chester
2974	Lord Barrymore	3/05	Bristol	8/33	Pontypool Road
2975	Lord Palmer	3/05	Bristol	11/44	Banbury
2976	Winterstoke	4/05	Bristol	1/34	Swindon
2977	Robertson	4/05	Paddington	2/35	Westbury
2978	Charles J. Hambro	4/05	Bristol	8/46	Swindon
2979	Quentin Durward	4/05	Bristol	1/51	Newport
2980	Coeur de Lion	5/05	Paddington	5/48	Gloucester
2981	Ivanhoe	6/05	Newton Abbot	3/51	Banbury
2982	Lalla Rookh	6/05	Plymouth	6/34	Newport
2983	Redgauntlet	7/05	Paddington	3/46	Tyseley
2984	Guy Mannering	7/05	Paddington	5/33	Pontypool Road
2985	Peveril of the Peak	7/05	Paddington	8/31	Landore
2986	Robin Hood	7/05	Paddington	11/32	Swindon
2987	Bride of Lammermoor	8/05	Exeter	10/49	Hereford
2988	Rob Roy	8/05	Newton Abbot	5/48	Tyseley
2989	Talisman	9/05	Bristol	9/48	Chester
2990	Waverley	9/05	Plymouth	1/39	Cardiff
2998	Ernest Cunard	3/03	Paddington	6/33	Cardiff

* 'Paddington' depot was in fact Westbourne Park later replaced by Old Oak Common.

† No 2925 *Saint Martin* – withdrawal date is that for rebuilding as prototype for Hall class 4-6-0s. See No 4900 for subsequent details.

CHAPTER THREE

MIXED TRAFFIC MOGULS

The Krugers

The most numerous 2-6-0 design built in this country was the Churchward 43XX class of 1910; although highly successful in this form many became part of later 4-6-0 types. They were not the Great Western's first class of 2-6-0, and were preceded by the ungainly looking Aberdare class of 1900. It is an irony to reflect that these 2-6-0s were themselves successors to the even more ungainly looking 'Kruger' and 'Mrs Kruger' designs, which were originally a 4-6-0 and 2-6-0, respectively.

In 1899 No 2601, a 4-6-0 with two inside cylinders, double frames, and 4ft 7½in coupled wheels, was built at Swindon Works Lot No 116 and works number 1723. It was followed by No 2602, to the same Lot No, but constructed as a 2-6-0, with the main frames cut away at the leading end as though for a 4-6-0 wheel arrangement. These were the first two of the Kruger class, which eventually came to number 10 locomotives in all, with running numbers from 2601 to 2610, with the last one coming out in 1903. Like the first GWR 4-6-0, No 36, these locomotives carried parallel boilers, with Belpaire pattern fireboxes. In the Krugers, this firebox was extended into a 3ft 6in long combustion chamber, shortening the boiler barrel proper to a length of 10ft 6in. Both Nos 2601 and 2602 carried a valve under the combustion chamber to provide a supplementary air supply. The maximum outside diameter of the boiler was 4ft 10in, and the grate area of 32.19sq ft was the largest included in any British locomotive of the time. Within the boiler 324 1⅞in diameter tubes provided over 1700sq ft of evaporative heating surface, while the firebox and combustion chamber contributed another 166sq ft to the total heating surface of 1879.68sq ft. A feature carried over from Dean's No 36 was the provision of large sandboxes mounted on the front boiler ring, rather like saddles. In 1903 and 1904 these were removed, and replaced on Nos 2601 and 2602 with steam sanding equipment and the sandboxes car-ried in more conventional positions.

It was in cylinder and valve design that G. J. Churchward's influence was most apparent in the first Krugers. The 19in × 28in cylinders were given 8½in diameter piston valves above the cylinders, operated by Stephenson valve gear. Three new ideas were introduced here: the long stroke of 28in, piston valves with a travel of 4½in, and single slide bars. The piston valves themselves were actuated indirectly by a system of rocking shafts.

No 2601 had a slightly longer wheelbase than the 2-6-0s at 25ft 0in, but was given the same diameter (2ft 8in) bogie wheels, and coupled wheels. The locomotive itself, at 63 tons 16 cwt, was 3 tons 8 cwt heavier, but was paired with a tender four tons lighter, resulting in all-up weight 17 cwt less than the 2-6-0. Main suspension was also different between Nos 2601 and 2602, with the 4-6-0 having coil springs on the main axleboxes; the bogie of No 2601 had inside frames similar to the prototype 4-6-0 No 36. In the first 2-6-0, No 2602, the coil springs on the leading coupled axle were replaced by laminated leaf springs. The basic dimensions of the Krugers are shown in the table overleaf.

Distribution of the locomotive's weight on the coupled axles was fairly even, although the 4-6-0 was markedly heavier on the driving and trailing axles. No 2601 lasted just over four years, and amassed a mileage of 67,000 miles in service. Although not a particularly outstanding total, it was still better than that achieved by the 2-6-0s Nos 2603–2610, whose poor average was between 30,000 and 50,000 miles.

The boilers fitted to the Krugers was designated type BR0X, and although quoted as 180 lb above, the working pressures varied between 200 lb/sq in and 165 lb/sq in, with three different sizes of boiler tube. Fourteen boilers were built to this same basic design between 1899 and 1904, having two rings, and with some influence from Churchward, they were domeless. Originally, No 2601's boiler was pressed to 200 lb/sq in, though this was later reduced to 180 lb/sq in, the same pressure carried by Nos 2602 and 2603. In subsequent members of the class it was reduced still further to 165 lb/sq in, with a corresponding reduction in tractive effort, from 30,960 lb with 200 lb/sq in boiler pressure to 25,543 lb at 165 lb/sq in pressure. Another experimental idea included on the Krugers was the fitting of corrugated wrapping plates to the combustion chambers. Despite some problems with these advanced, and fairly complicated designs of boiler for the period and their short working life on the locomotives themselves, some examples survived for many years as stationary steam plants.

Kruger class 2-6-0 – leading details

Wheel arrangement	2-6-0
Running Nos	2602–2610
Works Nos	1724–1732
Lot No	116
Built	6/1901–6/1903
Withdrawn	1/1906–1/1907
Wheel diameter – coupled	4ft 7½in
– pony truck	2ft 8in
Wheelbase	23ft 6in
Boiler – diameter	4ft 10in
– length	10ft 6in
– tubes	324 × 1⅞in o/d
– combustion chamber length	3ft 6in
Cylinders – bore × stroke	18in × 26in
Heating surface – tubes	1712.88 sq ft
– firebox/combustion chamber	166.80 sq ft
– total	1879.68 sq ft
Firebox length o/s	7ft 0in
Grate area	32.19 sq ft
Boiler pressure	180 lb/sq in
Tractive effort	27,865 lb
Weights – locomotive	60 tons 8 cwt
– tender	36 tons 15 cwt
– total	97 tons 03 cwt

There was also a proposal from Churchward in later years that the Krugers could be fitted with a taper boiler. A design was prepared with this in mind, but none was ever built. In addition to the troubles experienced with expansion in the original boiler design, the various types of piston valve used proved similarly unsatisfactory. Rebuilding the Krugers was finally ruled out with the success of the Aberdare design.

Operationally, as freight locomotives, the Krugers were employed on heavy coal workings between Swindon and South Wales, but performed very little useful work, as the low mileages at withdrawal would suggest. Even the highest aggregate mileage recorded by No 2602 revealed that the locomotive had covered an average of only 19,000 miles a year, with the average annual mileage of the remainder falling between 9,500 and just less than 16,000 miles a year. The working lives of Nos 2603–2610 was a little over three years, with those of Nos 2601 and 2602 only slightly longer at five and six years respectively. The construction and withdrawal dates are as shown in the table:

Running No	Built	Withdrawn	Mileage worked
2601	12/1899	12/1904	67,000
2602	6/1901	1/1907	80,000
2603	1/1903	2/1906	
2604	1/1903	4/1906	
2605	1/1903	2/1906	Average
2606	2/1903	5/1906	30,000
2607	2/1903	1/1906	to
2608	3/1903	12/1906	50,000
2609	3/1903	12/1906	
2610	6/1903	8/1906	

The Aberdare Class 2-6-0s

The 81 locomotives which eventually came to represent the 2-6-0 Aberdare class derived originally from locomotive No 33, built in 1900 to Lot No 128. In service as a class, they carried running numbers 2600 to 2680, and it is reported incorporating for some at least parts from the earlier Kruger class, in addition to their old running numbers. The Aberdares were the goods versions of the Bulldog and Atbara class 4-4-0s, and were much more attractive looking than their immediate predecessors. They were a successful and long-lived design – intended like the Krugers for working coal trains between South Wales, Aberdare in particular, and Swindon – with a dozen of their number lasting into British Railways days.

The pioneer Aberdare No 33 began life with a parallel boiler, but a raised top Belpaire type firebox, piston valves, and no superheater. Similarly, another 40 locomotives numbered between 2621 and 2660, were outshopped from Swindon in like manner, to Lot Nos 131 and 133, in 1901

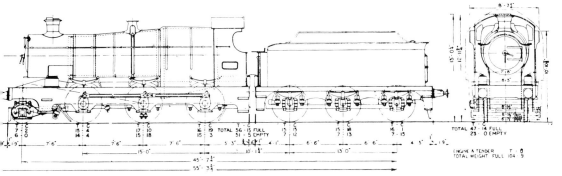

Aberdare class 2-6-0

Below. The original parallel boilered Aberdare 2-6-0, a very rare view showing tall safety valve covers, no top feed, tall narrow chimney, and coal rails on tender. The train is running over longitudinal sleepered track. *Locomotive & General Railway Photographs*

Bottom. Typical of an Aberdare in later life, No 2602 has top feed, a tapered boiler, and chimney with capuchon. *Locomotive & General Railway Photographs*

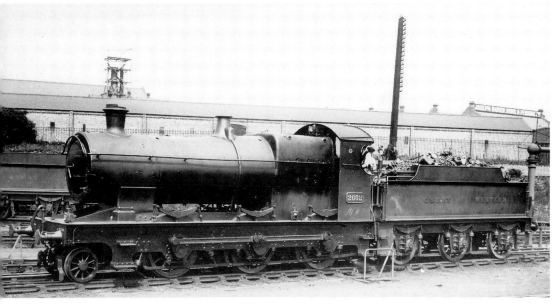

and 1902. These were fitted with the standard No 2 parallel boiler, and it was not until No 2662 appeared in September 1902, with a long taper standard No 3 boiler that the appearance of the Aberdares took on the more familiar form. The accompanying table gives the leading details of the class, fitted with the parallel, Standard No 2, and tapered Standard No 3 boilers:

Aberdare class 2-6-0 – leading details

Wheel arrangement	2–6–0
Running numbers	2600–2680
Works numbers	1796–1805, 1886, 1908–1968, 2125
Swindon Lot numbers	123, 128, 131, 133, 135, 156, 166
Built	8/1900–12/1906
Withdrawn	8/1934–10/1949

Wheel diameter	
– coupled	4ft 7½in
– pony truck	2ft 8in
Wheelbase	22ft 6in

Cylinders bore × stroke	18in × 26in

Boiler	*Taper boiler*	*Parallel boiler*
– diameter	4ft 10¾in/5ft 6in	4ft 6in
– length	11ft 0in	11ft 0in
– tubes – number	350	277
– diameter	1⅝in o/d	1⅝in o/d
Heating surface		
– tubes	1689.82 sq ft	1538.06 sq ft
– firebox	128.30 sq ft	124.96 sq ft
– total	1818.12 sq ft	1663.02 sq ft
Firebox		
– length	7ft 0in	7ft 0in
– width	5ft 0in	5ft 0in
Grate area	20.56 sq ft	21.45 sq ft

Weights *Taper boiler*	Tons	Cwt
– Pony truck	7	2
– Leading	15	4
– Driving	17	10
– Trailing	16	19
– Total	56	15

Parallel boiler	Tons	Cwt
– Pony truck	6	12
– Leading	15	16
– Driving	15	10
– Trailing	15	10
– Total	53	8

Tractive effort	25,800 lb (taper boiler)
	23,322 lb (parallel boiler)

Tender	
– weight	36 tons 15 cwt
– water capacity	3,000 gallons

Boiler pressure	200 lb/sq in (taper)*
	180 lb/sq in (parallel)

* Some locomotives nominally 200 lb/sq in had boilers pressed to work at 195 lb/sq in only.
NB: The figures for total heating surface quoted in this table do not take account of later alterations brought about through the fitting of superheaters.

Construction details

The Standard No 2 parallel boilers fitted to No 33 and Nos 2621–2660 were domeless, and built with two rings. One of these boilers was fitted with the 2½in diameter internally ribbed *Serve* tubes, although by far the majority had plain bore 1⅞ od tubes, with minor dimensional differences in both boiler and firebox assemblies. No 33, renumbered 2600 in December 1912, sported other distinguishing features, such as the vertical spring hangers on the leading coupled axleboxes, with a large compensating beam between the pony truck and the leading axle.

Before the parallel boiler Aberdares were finally rebuilt with taper boilers, a number had already been at work with different boilers, having a taper on the rear section only. Nos 2635, 2649, and 2655 were subjected to these changes in April and May 1904, while all the members of the 2621–2660 series were rebuilt with Standard No 4 taper boilers between January 1903 and April 1910. Most of the class were reboilered in 1903. Originally Nos 33 (alias No 2600), and Nos 2621–2660 were covered by two diagrams, B and C respectively, with saturated boilers and 6½in diameter piston valves carried below the cylinders. No 33 retained its piston valves until new cylinders were fitted in October 1904, with slide valves, which in their turn lasted until 1915, when piston valves were fitted once again.

Other experimental features fitted to the second batch of Aberdares, Nos 2621 to 2640, included steel fireboxes, trick-port piston valves, and both long and short taper boilers. No 2661 built in September 1902 was the only member of the class turned out new with a Standard No 4 parallel boiler, which was removed two years later when a conventional tapered No 4 boiler was fitted.

The first to be built in what came to be the Aberdare's standard form also arrived in September 1902, was No 2662 with Standard No 4 boiler. Boilers for these locomotives in their standard form included three minor variations, chiefly concerning the number of tubes. The first 20 boilers carried only 312 tubes of 1¾in diameter, while the majority housed 350 of 1⅝in diameter, with an evaporative heating surface of 1689.82sq ft. Five boilers were constructed in 1905 to the same basic format, but with the internal dimensions reduced by 1in in

length and width, reducing its heating surface slightly, while that provided by the tubes increased to 1708.04sq ft.

The standard locomotives were also fitted with slide valves from new, although they were all equipped with piston valves from 1915 onwards. One of the reasons for the reintroduction of slide valves with the Aberdares was the failure of Churchward's early design of piston valve. The 18in × 26in cylinders were set between the frames, supported the drumhead type smokebox on a fabricated saddle. Main axleboxes were plain bearing, with flycranks driving the fluted coupling rods outside the frames, and overhung leaf springs to each axle. The spring hangers on the leading axle were inclined noticeably with the front hanger extending to reach below the lowest step on the front footsteps. Brakes were carried in single hangers, ahead of the coupled wheels, with sanding to the front of the leading and in the rear of the trailing coupled wheels.

There were a number of variations in boiler fittings, with some locomotives having top feed apparatus, and others with the short style of safety valve bonnet, while three basic designs of chimney could be seen. The original type was cast-iron, and reboilerings of individual locomotives resulted in the occasional appearance of City class chimneys, with the copper cap. One unusual arrangement to replace the original cast-iron chimney, included a tapered lower

Right-hand side view of No 2622 in 1935 showing short pattern safety valve and top feed, tapered chimney without capuchon, lubricator pipe covers, and snap-head rivetting around boiler-smokebox joint.
HMRS/L. E. Copeland

half with parallel copper cap, while a second design was tapered throughout. In later years, a new pattern of tapered cast-iron chimney was applied.

Top feed apparatus did not appear until March 1911 when it was first fitted to No 2625 on the front boiler ring between the chimney and safety valves. Most of the class, including the replacements for the Krugers, also began to receive superheaters from about this period. However, a year or two earlier, in December 1908, a 72-element superheater (classified as Swindon No 2) was fitted to No 2679, and provided 215.0sq ft of heating surface. This was a three-row design, and unique to this locomotive, since the rest of the class were provided with the more traditional two-row superheaters. These later fittings of two rows, with fourteen $5\frac{1}{8}$ diameter flues contained 84 elements, although between 1914 and 1921 some Aberdares carried superheaters with 112 elements, providing a heating surface of 250.0sq ft.

Tenders attached to the Aberdares, including the earlier prototype locomotives, had a water capacity of 3,000 gallons, and a wheelbase of 15ft 0in, equally divided. Similar tenders were paired with Dean's Bulldog

The left, fireman's, side view of No 2643 shows the taper on the rear boiler ring, and the inclination and length of the spring hangers for the leading axle.
HMRS/L. E. Copeland

Diagram	Running Nos	Variations/features
B	2600	Standard No 2 boiler, piston valves
C	2621–2660	Standard No 2 boiler, piston valves
D	2661–2680	Standard No 4 boiler, slide valves
E	2679	Standard No 4 boiler, 3-row superheater, slide valves
F	26XX*	Standard No 4 boiler, 14 flue/84 element superheater, slide valves
I	26XX*	Standard No 4 boiler, 14 flue/112 element superheater, piston valves
Q	26XX*	Standard No 4 boiler, 14 flue/84 element superheater, piston valves and ROD tender

* Basic Aberdare class 2-6-0

and Duke class 4-4-0s, and the 0-6-0 Dean Goods. Initially, only coal rails were provided, but at a later date these were replaced by sheet steel fenders. After 1929, fifty ROD tenders with a water capacity of 4,000 gallons were paired with various members of the class, as the former ROD 2-8-0s were scrapped. Only 20 Aberdares were known not to have been paired with this tender during their working lives. It was also necessary to equip the locomotives with vacuum brake apparatus.

Operations
The Aberdares were freight locomotives, employed on coal workings initially, between South Wales and Swindon. Seven diagrams were issued covering the major mechanical variations within the class as shown in the table on the right.

Naturally, many were shedded in South Wales, at Aberdare, Pontypool Road, and other locations, with others in the Midlands, at Tyseley, Oxford, and Croes Newydd. Allocations were also made to Swindon and Paddington. After World War I, most could be found in the Newport area, working to Oxford, the Midlands and Birkenhead, with Cardiff based locomotives working to

Exeter, and those allocated to Newport and Severn Tunnel Junction running through to London. In Devon and Cornwall mineral traffic was their principal load, with similar work in other divisions. As the big 28XX 2-8-0 locomotives took on responsibility for the heaviest coal trains in their later years, the Aberdares were most often seen on local freight and mineral trains, concentrated in the Worcester and Wolverhampton divisions, and in South Wales.

Withdrawals began in the mid-1930s, continuing until 1938 at the rate of approximately one a month, while in 1939 five were taken out of service, stored, and subsequently re-instated for wartime traffic.

Withdrawals began again in 1944, with the last locomotive taken out of service in October 1949 as British Railways No 2667. This locomotive had covered 1,071,953 miles since entering service in August 1903 at St Blazey in Cornwall. Almost a quarter of the class had run more than a million miles, with another 31 passing the ¾-million mark.

Of the 11 Aberdares still in service in 1948, three had not covered more than a million miles, and at the end of 1947, the remaining locomotives were to be found at the following depots:

Running No	Last GWR depot	Miles run	Date withdrawn
2612	Banbury	905,127	1/1948
2620	Stourbridge	1,048,849	8/1949
2643	Banbury	1,070,658	7/1948
2651	Gloucester	1,049,082	6/1949
2655	Stourbridge	995,553	6/1949
2656	Gloucester	1,050,226	3/1948
2662	Chester	1,076,973	7/1948
2665	Oxley	1,003,452	1/1948
2667	Pontypool Road	1,071,953	10/1949
2669	Pontypool Road	1,015,482	5/1948
2680	Hereford	910,434	6/1948

In their final couple of years, the Aberdares were mainly employed in the Wolverhampton and Worcester districts, with two in the Newport area. They had been employed on local duties in these areas since about 1935. As British Railways locomotives they were power class 4F, but none was known to have carried either a smokebox numberplate, or the temporary W prefix used in 1948. No 2669 at Pontypool Road

was paired with an ROD tender. No 2662 engine held the distinction of being the first standard Aberdare, appearing in September 1902 with the tapered No 4 boiler.

The 43XX class moguls

This immensely successful Churchward designed 2-6-0 first appeared in 1911, with its two outside cylinders and inside Stephenson valve gear. It was produced as a result of Churchward's desire to have a general-purpose locomotive in the standard range then being developed for the GWR. Its appeal was further enhanced following a visit by Harry Holcroft to the USA, who upon his return to Swindon produced an impressive report of the 2-6-0 moguls' operations in that country.

The new Churchward moguls were the tender equivalent of the 3150 class 2-6-2 tank engines, and utilised many standard components. Ultimately, no fewer than 342 were built over a 21-year period between 1911 and 1932; not all were built at Swindon, however, with Robert Stephenson & Hawthorn supplying 35 in 1921/22. Almost

43XX class 2-6-0

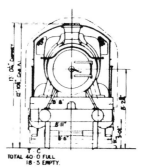

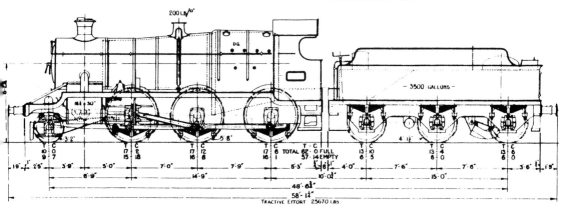

One of Churchward's most numerous and successful designs was the 43XX 2-6-0, produced from 1911 onwards, with this example piloting a King class 4-6-0 on a passenger train on one of the arduous banks in South Devon. Note the tall safety valve cover. *Lens of Sutton*

weight distribution. Tenders paired with the moguls carried 3,500 gallons of water and 7 tons of coal, with a weight in working order of just 40 tons.

one-third of the class was built during World War I, and proved to be just the locomotive that the running department needed at that time. With so many locomotives being built over such a long period there were inevitably numerous variations. Having said that, there is a rumour which persists that one GWR locomotive type was built from only 11 new drawings, all the remaining details being produced from existing drawings – the 43XX 2-6-0 may well have been that locomotive. They were undoubtedly maids of all work, and a true mixed-traffic type, working semi-fast trains, the longer branch lines, and cross-country services. The 5ft 8in diameter coupled wheels, adopted as a standard size were first seen on Dean's mixed traffic design, the Duke or Devon class 4-4-0. Churchward's moguls were the first new locomotives to come out of Swindon with top feed apparatus from new, while the earlier locomotives of the first batch up to No 4309, were the last to carry Swindon Works plates.

Twelve diagrams were issued in all covering the 43XX 2-6-0s, including such variations as longer cabs, internal alterations to boilers, outside steam pipes, oil-burning equipment, and a number with altered

Diagram	Running Nos	Features/variations
G	4301–4320	Standard locomotives
H	4321–4396	Standard locomotives
J	4397–4399, 4300	Altered weight distribution
K	5300–5399 6300–6341 7300–7319	Original condition
	4300, 4321–4399	Altered weight distribution
L	4301–4320	Altered weight distribution
	6342–6369 7320–7321	As built, earlier locomotives were fitted with 84-element superheaters, in new boilers built between 1923 and 1940. Outside steampipes are also shown on this diagram
P	83XX	65 locomotives with altered weight distribution
R	9300–9319	Standard
S	6320	Fitted with oil-burning equipment
U	9300–9319	Modified 1956–59 and renumbered 7322–7341
V	All except 7322–7341	Locomotives with altered tube arrangements in new boilers
W	7322–7341	As diagram V

Power classification: D – all locomotives.
Route restrictions: Blue – 43XX, 53XX, 63XX, 73XX
Red – 83XX, 93XX.

Construction

The boilers fitted to 43XX moguls were Standard No 4, of which 744 were constructed between 1911 and 1940, with seven distinct variations. The majority housed 235 small tubes, with the attendant increase in the heating surface to 1349.64sq ft. A similar increase in the superheater heating surface resulted from the addition of a further 38 elements in the flues. The total evaporative heating surface of the boilers had been increased by almost 9½%, and yet the Standard No 4 boilers built between 1923 and 1940 reverted to the original 84-element superheaters. The boilers fitted originally to Nos 4311–4320 were designed to operate at 225 lb/sq in pressure, although only Nos 4315 and 4316 are believed to have worked at that pressure.

The 7ft 0in firebox was standard GWR Belpaire topped design, waisted-in to fit between the locomotive mainframes, and while the external dimensions remained unaltered there was some variation in the internal length and width. The majority were 6ft 2½in long and 4ft 8⅞in wide at the leading end, and 3ft 2⅝in at the rear. Sloping upwards and rearwards to the firehole the internal height varied from 6ft 6⅜in at the front, to 5ft 0⅜ at the back.

The now standard cylindrical smokebox was carried on a saddle, half of which was cast integral with the left-hand cylinder assembly, and half with the right-hand, in another Churchward modification of contemporary American practice. Bar frame extensions were bolted onto the main plate frames which terminated just behind the cylinders, running under the cylinders to the front buffer beam, supporting the front footplate and pony truck suspension. In the type of bar frame construction adopted for the smaller wheeled Churchward designs, the relative weakness of the front end was obviated by the fitting of steel struts attached to the smokebox saddle and the front end of the bar frame extensions. The 18½in diameter cylinders were horizontal, and as by then standard practice the cylinders were on the same centre line as the coupled axleboxes. Only the coupling, connecting rod and crosshead were outside the frames, with Stephenson valve gear inside. An extension of 9in to the rear of the main plate frames on locomotives built after No 4320 gave more room for pipework, and resulted in the appearance of longer cabs.

The first 320 or so were more-or-less standard, with only minor variations in details, and it was not until late 1927 that any major changes were made. In 1927 and 1928, to counter the effects of flange wear on the leading coupled wheels, which was particularly marked on those locomotives working in the West Country and Cornwall, the locomotive weight distribution was altered. The first four so treated, Nos 4351/86/95/85, were renumbered 8300/35/44/34, although they very quickly reverted to their original numbers. Between January and March 1928, another 65 of the 53XX series were modified and given numbers in the 83XX series. The majority of these 65 locomotives, which were restricted to 'red' routes, lasted in this form until around 1944, although a dozen had been cannibalised earlier to provide parts used in the building of Grange class 4-6-0s. The remaining fifty-three 83XX 2-6-0s had their extra weights removed and were restored to their original condition by 1948, replacing withdrawn 4-4-0s operating over 'blue' routes.

New cylinders were fitted to 163 locomotives between July 1928 and August 1958, with outside steampipes. Ten locomotives were built in this form in 1925, before the 93XX series emerged as the 'top of the range' model in 1932. Out of the 241 locomotives which survived into British Railways ownership, most appeared in this form. The twenty 93XX series locomotives included modifications introduced by C. B. Collett, such as the cab with side windows, screw reverse. Rather surprisingly, they came into service without the GWR AWS apparatus. The 93XX series locomotives were restricted to 'red' routes, like the 83XX series, due to the additional weights fitted at the front end to improve adhesion. This cast weight was carried immediately behind the front buffer beam, making it appear much thicker than on a conventional 43XX mogul. In later years, the 93XX locomotives under British Railways ownership underwent modification, with the removal of this additional weight, and were renumbered in the 73XX series. While most of the 93XX series were fitted with short safety valve bonnets, some had the taller type fitted to the majority of the standard moguls. (The short pattern

This excellent study of a Churchward mogul, shows No 5341 in British Railways fully lined green livery, at Tyseley shed in 1958. *L. C. Jacks*

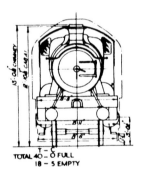

TOTAL 40 – 0 FULL
18 – 5 EMPTY

ENGINE & TENDER   T – C
TOTAL WEIGHT FULL 103 – 17

The later GW 2-6-0s of class 7322

safety valve bonnets began to appear on the standard locomotives from around 1926.)

Detail variations in this design ranged from the positioning of the lamp brackets, through cylinder layout, type of reversing gear, the fitting of AWS equipment, and automatic tablet catchers on the tenders of some locomotives working the North Devon branches. AWS, or as it was less correctly known, Automatic Train Control (ATC), was fitted to seven locomotives in 1928, with the remainder acquiring the apparatus between January 1930 and April 1932. From the 1930s, a number of locomotives were paired with a Collett design of 3,500-gallon tender. Nos 6332 and 6336 were equipped for a short time in the 1930s with Westinghouse air brake equipment, while another unusual acquisition was an all-welded

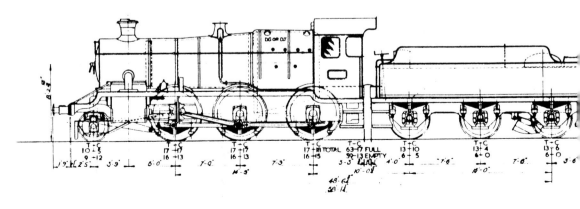

copper firebox carried by No 4371 from 1933 until its withdrawal in 1938.

After World War II conversion of GWR locomotives to oil burning was pursued with some enthusiasm for a time, and in 1947 No 6320 was equipped for this. The oil burning equipment installed raised the mogul's weight in working order to 62 tons 6 cwt, and although only one mogul was altered, it is possible that No 9314 was also scheduled for conversion. This carried the cab side window shutters, in common with other oil burning locomotives, but it was never fully modified.

The most widely known 'conversion' of the Churchward moguls was the use of wheels, motion, and some other components in the new Collett Grange and Manor class 4-6-0s. Essentially, the 100 locomotives withdrawn in the mid to late 1930s were converted to Grange and Manor class locomotives; although it was planned to convert all the 2-6-0s in this way, the programme was never resumed after the hostilities had ended. Eighty-eight of the conversions came from the 43XX series, and 12 from the 83XX, 'red' route locomotives.

Operations

From their introduction in Churchward's day until well into the 1930s the moguls were handling a very wide variety of traffic over a varied range of routes. Anything from fast freights to express passenger services was within their capabilities, and while the legendary story of a 53XX on a freight overtaking the Cheltenham Flyer is well known, No 6307 deputised for a failed diesel on the Cornish Riviera Express in BR days. During World War I Nos 5319–5326 and 5328–5330 were sent to France in the service of the Railway Operating Division of the British army.

The 83XX and 93XX with the heavier front end loading were distributed mainly throughout Devon and Cornwall, with allocations also to Wolverhampton, Bristol, Newport and the London area. Prior to World War II, none of the 'blue' route locomotives was allocated to the Central Wales Division, and most were stabled in the Wolverhampton and Bristol areas. There was a fairly even distribution of the remainder between London, Worcester, Newport and Neath divisions, with a smaller number in the West of England. When certain 'yellow'

43XX class 2-6-0 – leading details

Wheel arrangement	2-6-0		
Running numbers	4300–4399, 5300–5399, 6300–99, 7300–21, 9300–19		
Works Nos	2396–415, 2516–35, 2552–71, 2612–31, 2652–2761, 2790–2841, 2887–2906, 3802–36		
Lot Nos	183/193/194/198/202/204–209/211/212/216/218/222/230/276		
Built	6/1911–4/1932		
Withdrawn	2/1936–11/1964		
Wheel diameter			
– coupled	5ft 8in		
– pony truck	3ft 2in		
Wheelbase	23ft 6in		

Axle load	Tons Cwt leading	Tons Cwt driving	Tons Cwt trailing
– coupled	18 04	17 12	16 16
– pony truck	9 08		

Boiler		
– diameter	4ft 10¾in, tapering to 5ft 6in	
– length	11ft 0in	
– tubes, small	209×1⅝in (235×1⅝in from No 4321 onwards)*	
large	14×5⅛in	
– superheater elements	84×1in o/d	

Cylinders – bore × stroke	18½in × 30in

Heating surface		
– tubes	1228.02 sq ft	(1349.64 sq ft)*
– firebox	122.92 sq ft	128.72 sq ft)*
– total	1350.94 sq ft	(1478.36 sq ft)*
– superheater	215.80 sq ft	(212.58 sq ft)*

Firebox – length, o/s	7ft 0in

Grate area	20.56 sq ft
Boiler pressure	200 lb/sq in

Tractive effort	25,670 lb

Fuel capacities	
– coal	7 tons
– water	3,500 gallons

Weights	Tons Cwt
– locomotive	62 00
– tender	40 00
– total	102 00

* These figures refer to the second batch, Nos 4321–40, built in 1913, with the major dimensions common to all subsequent builds. The frame extension – 9in longer to the rear – together with these alterations produced variations in axle loading on the original 43XX 2-6-0.

routes were upgraded to 'blue' around the time of World War II, the heavier moguls then found their way into Central Wales, although some from the Chester and Croes Newydd allocations had already worked over these lines.

Further re-allocations were made during BR days, although only 241 locomotives were taken into British Railways stock. One hundred had been withdrawn to provide components for the Grange and Manor class 4-6-0s. No 6315 had been scrapped following an accident on 7 September 1944, when a breach in the Shropshire Union Canal caused the railway embankment near Llangollen to be washed out. The locomotive had been hauling the Chester to Barmouth mail train when it and the first two coaches plunged into the washout near Sunbank Halt on the Ruabon side of Llangollen.

In 1948 No 5345/88 and 6358 were given the W prefix, under the temporary numbering scheme adopted by the new organisation. Eleven locomotives did not receive the new cast smokebox numberplates. 43XX moguls were widely distributed in the 1950s, although many had been withdrawn, replaced on light passenger duties by new diesel multiple-unit trains. By 1957, only 210 were still in service, but less than seven years later all had been scrapped.

The moguls were most frequently seen on freight and local passenger workings at the end of their lives – almost exactly the same duties they were engaged on at their introduction almost half a century earlier. They were also to be found occasionally at the head of express passenger trains, particularly during the summer peak periods. Their last year of service, 1964, saw only 35 survivors, at Didcot, Aberdare, Gloucester, Leamington, Neath, Pontypool Road, Severn Tunnel Junction, Stourbridge and Taunton. Even to the last, a 43XX 2-6-0 could be found requisitioned for an unusual tour of duty – Nos 6372/85 were immaculately turned out to work the Royal Train from Taunton to Barnstaple in May 1956. Not surprisingly, many of this class had recorded more than a million miles in service by the time of their withdrawal.

Under the Great Western classification system, the majority of the moguls were power class D, and could operate on 'blue' routes, while the heavier weighted engines in the 83XX and 93XX series were restricted to 'red' routes, but were still power class D. As British Railways locomotives, they were all power class 4MT.

In 1950, there remained 230 of these locomotives, allocated to the following districts, thus: Newport 11, London 45, Wolverhampton 53, Neath 22, Bristol 40, Worcester 28, Newton Abbot 26, Birkenhead 5.

The moguls continued to be used on their traditional workings in BR days, with some inter-regional running such as Reading to Redhill, Didcot to Southampton, and other routes. They were to be found on all varieties of passenger and freight traffic – truly mixed traffic engines. Early in their British Railways service, like the Grange and Manor classes, some of the moguls had their blastpipe jumper rings removed. Once again, this alteration had beneficial results on their performance. A final modification was carried out to these 2-6-0s between 1956–1959, when the 93XX series engines had the heavy cast metal weights (carried behind the front bufferbeam) removed, and were renumbered 7322–7341. At this time, a new diagram U was issued, covering the modified 2-6-0s. No 6320, which had been fitted for oil burning in 1947, had the equipment removed in August 1949. In the late 1950s fifteen new boilers were built at Swindon for the moguls, with three row superheaters, and fitted to a number of locomotives, resulting in the appearance of two more diagrams, V and W.

In the early BR period, the moguls were classed as mixed traffic but carried only plain black livery, with No 7313 being given mixed traffic black, lined red, cream and grey. In common with a number of other ex-GWR types in the late 1950s, green livery was again widely used, later still with full lining. It was not, however, carried by all the 43XX moguls which remained in service in British Railways ownership.

THE HALL CLASS 4-6-0s

This class of 4-6-0 was easily the most numerous on the GWR, and was a design whose ancestry can be directly traced to the Saint class. In a number of instances they have been referred to as '6ft 0in Saints'. No fewer than 258 of Collett's Hall class locomotives were built between 1928 and 1943, following the successful modification and operation of Saint class No 2925 *Saint Martin*.

Designs for the new Hall class locomotives were born out of Churchward practice, with some influence from the operating department. They were perhaps the first truly mixed traffic type owned and operated by the GWR, and although built when C. B. Collett was Chief Mechanical Engineer, the basic format had already been outlined by Churchward in his scheme of 1901. The success of the 43XX moguls would, said the running department, be improved still further with the lengthening of the wheelbase, the provision of a leading bogie, and for greater power a Standard No. 1 boiler.

In December 1924 No 2925 *Saint Martin* came out of Swindon Works with 6ft 0in coupled wheels and the Collett side window cab. But these were the most obvious dif-

ferences, with others thought necessary on the rebuild but not included later or modified for the production series locomotives. The rebuilding of *Saint Martin* incorporated standard Saint cylinders, which conventional Swindon practice required to be carried lower in the frames in order to line up with the centre line of the smaller coupled wheels. Since the half saddles were cast integrally with the cylinders, the boiler was 4¼in lower with its centre line pitched at 8ft 1¾in above rail level. This change naturally affected other components, of which another obvious example was the smaller coupled wheel splasher required, with the result that the original nameplates could not be refitted. New nameplates were produced, with the lettering more closely spaced. This change in style was a feature of the Collett era, and was applied to other locomotive designs produced under his direction.

This is the locomotive which gave birth to the most numerous, popular, and successful of the GWR's two-cylinder classes. The extensive modifications to No 2925 *Saint Martin*, here seen in original coverted form with inside steampipes, led to the building of the Hall class 4-6-0s. *Lens of Sutton*

Rebuilt Saint class 4-6-0 – leading details

Wheel arrangement	4-6-0
Running number	2925 (later 4900)
Works No	2273 (as Saint class locomotive)
Built	9/1907
Rebuilt as prototype Hall	12/1924
Date withdrawn	4/1959
Lot No	170 (as Saint class locomotive)

Wheel diameter	
– coupled	6ft 0in
– bogie	3ft 2in
Wheelbase	27ft 1in

Axle load	Tons Cwt leading	Tons Cwt driving	Tons Cwt trailing
– coupled	18 00	18 09	18 09
	Tons Cwt		
– bogie	17 12		

Boiler	
– diameter	4ft 10¾in tapering to 5ft 6in
– length	14ft 10in
– tubes, small	176 × 2in o/d
– tubes, large	14 × 5⅛in o/d
– superheater elements	84 × 1in o/d

Cylinders – bore × stroke	18½in × 30in

Heating surface	
– tubes	1686.60 sq ft
– firebox	154.78 sq ft
– total	1841.38 sq ft
– superheater	262.62 sq ft

Firebox – length, o/s	9ft 0in

Grate area	27.07sq ft

Boiler pressure	225 lb/sq in

Tractive effort	27,275 lb

Fuel capacities – coal	7 tons
– water	3,500 gallons

Weights	Full Tons Cwt	Empty Tons Cwt
– locomotive	72 10	66 00
– tender	40 00	18 05
– total	112 10	84 05

In the rebuilt locomotive, the reversing gear was changed, replacing the wheel-and-screw assembly of the type fitted to Star class locomotives with the arrangement provided on the Castle class. Some members of the Saint class – the Lady series and former Atlantics – were still fitted with lever reverse, even when Churchward was in the driving seat, but *Saint Martin* was one of those built with screw reverse. The rebuild, renumbered 4900 in December 1928 was tested over a period of four years, but even before that period of testing had expired the locomotive had proved so successful that an

order for a batch of 80 new locomotives to this basic design was placed on Swindon works on 23 December 1927.

The axle loading and other weights quoted for the rebuilt No 2925 produced a locomotive ½-ton heavier than the original Saint design, despite the considerable saving in weight made possible by the use of smaller diameter wheels. However, the diagrams for the original Saint class were issued in 1907, and did not include the considerable weight of the GWR audible warning equipment, notably the huge ramp contact shoe at the leading end of the locomotive. The inclusion of the weight of this equipment in the rebuilt *Saint Martin* may be the only area where an increase in weight over the original design could have been achieved.

As a member of the Hall class, No 4900 remained the only example with inside steam pipes, until 1948. In December that year, a new front end was fitted with new cylinders and outside steampipes, rendering the pioneer Hall class locomotive almost indistinguishable in some respects from the rest of the class.

The appearance of these new mixed-traffic locomotives was not without its troubles, despite the successful trials with the rebuilt *Saint Martin*, although none of these related to the design, construction, or operation of the new arrivals. On the GWR, as on other companies' lines, the 1930s were times when many of the older designs were being scrapped and replaced with more

modern, more efficient locomotives. However, the rail enthusiasts of that time regretted the arrival of the Hall class, if only because many of the ageing 4-4-0s – some dating back to the 1890s – would soon be extinct.

The first 80 of the new breed came out of Swindon Works under Lot No 254 between December 1928 and February 1930. A number of differences with the rebuild prototype were immediately apparent in the appearance of this first production series, including the raising of the pitch of the boiler to 8ft 6in, back to where it was in the Saint class. Outside steampipes were

Prototype for the Hall class, No 4900 *Saint Martin*, as rebuilt in 1948 with outside steampipes. Compare with the photograph on page 37. *A. R. Brown*

another identifying feature, while some changes had also taken place in the layout of the valve gear. Initially, 3,500-gallon Churchward tenders were paired with the new locomotives, but this was altered after No 4942 as the following 15 locomotives had a newer Collett-designed 3,500-gallon ca-

The first Halls, of which No 4909 *Blakesley Hall* was one, were built to Lot No 254, with differences between the new locomotive and the rebuilt *Saint Martin* including outside steam pipes and higher pitched boiler. *Lens of Sutton*

pacity tenders. A third type of tender most commonly paired with Halls was the 4,000-gallon Collet version.

Hall class 4-6-0 – leading details

Wheel arrangement	4-6-0
Running numbers	4901–99, 5900–99, 6900–58
Lot numbers	254, 268, 275, 281, 290, 297, 304, 311, 327, 333, 338, 340, 350, 366, 368, 376
Built	12/1928–4/1943
Withdrawn	11/1959–12/1965

Wheel diameter	
– coupled	6ft 0in
– bogie	3ft 0in
Wheelbase	27ft 1in

Axle load	Tons Cwt leading	Tons Cwt driving	Tons Cwt trailing
– coupled	18 12	18 19	18 19
	Tons Cwt		
– bogie	18 10		

Boiler	
– diameter	4ft 10¾in tapering to 5ft 6in
– length	14ft 10in
– tubes, small	176 × 2in o/d
– tubes, large	14 × 5⅛in o/d
– superheater elements	84 × 1in o/d

Cylinders	
– bore × stroke	18½in × 30in

Heating surface – tubes	1686.60 sq ft
– firebox	154.78 sq ft
– total	1841.38 sq ft
– superheater	262.62 sq ft

Firebox length o/s	9ft 0in

Grate area	27.07 sq ft

Boiler pressure	225 lb/sq in

Tractive effort	27,275 lb

Fuel capacities – coal	7 tons
– water	4,000 gallons

Weights	Full Tons Cwt	Empty Tons Cwt
– locomotive	75 00	69 00
– tender	46 14	22 10
– total	121 14	91 10

Boiler design

The boilers fitted to the new Hall class locomotives were the familiar Standard No 1, as adopted for all large ten-wheeled locomotives, and carried by their predecessors the Saint class. They were also given an index letter, in this case A, as part of a very detailed classification scheme brought into use on the Great Western between 1908 and 1912. The basic idea behind this was to accommodate variations in design within a standard class or group, and for the class A designs used on Halls, Saints, Stars, etc., there were 10 sub-divisions. Each of these sub-divisions was indicated in the second letter of the two letter index code, AF, AH, etc.

Hall class boilers were built in two rings, both of which were tapered, and with the Saint class, were from Standard No 1 boilers with index codes, AF, AH, AK, and AL. The conventional trapezoidal fireboxes were 9ft 0in long and had a maximum width of 5ft 9in over the wrapper plates, waisted-in to fit between the locomotive mainframes, where the minimum width was 4ft 0in. The firegrate itself was flat at the rear, but sloped downwards to the front of the box, from just in front of the trailing coupled axle. In heating surface, flues, tubes, and other major dimensions and components, the Hall class boilers were identical with the later builds of the Saint class.

Cylinders were 18½in diameter by 30in stroke, as standard for the two-cylinder types, but in this case the 10in diameter piston valves had had an increase in maximum travel to 7¼in, with inside admission. Stephenson valve gear was now standard, with the movement transferred through the frames by means of a rocking shaft from the extension rod. A new rocking shaft arrangement was provided on the production series Halls, while Nos 4901–4940, and 4951–4977 were turned out with valve spindle guides of the same design used on Saint class 4-6-0 No 2974 *Lord Barrymore*. This experimental arrangement lasted from 1928 until about 1936, with the outward sign being circular covers over the valve spindle guides, projecting over the front footplate. In the standard valves, the ports themselves were 31½in long, with a negative lead (−0.19in) in full gear, altering to a positive value (+0.27in) as mid-gear was approached. Steam lap was fixed at just under 1¾in (1.63in). This particular arrangement was also adopted on the Saint, 43XX, later Modified Hall, Grange, Manor, and County classes.

Frames, Wheels and Motion

While the coupled wheels of 6ft 0in diameter were the same as those fitted to No 2925 *Saint Martin*, the bogie wheels were reduced to 3ft 0in diameter. The front end frames were also modified in the production series so that the horizontal centre line of the cylin-

An attractive view of No 4936, *Kinlet Hall*, representative of the most common form of the class, paired with a Collett 4,000-gallon tender. *G. W. Sharpe*

This close-up view of slidebars and cylinders on No 6904, one of the later series, shows the long valve rod, back-pressure valve, piston gland, slipper block, and crosshead cotter securing piston rod to crosshead, sandbox, slidebars, and other details. *L. C. Jacks*

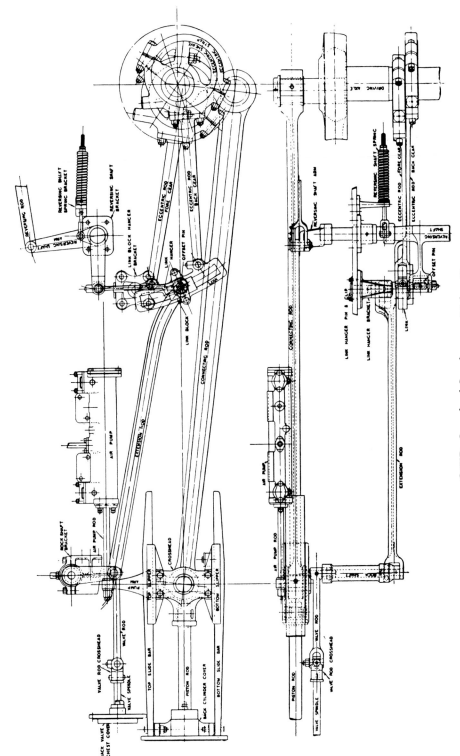

GWR. Standard Stephenson Valve Gear.

ders could be aligned with the centre line of the coupled axles. The new standard Churchward/DeGlehn type bar frame bogie continued to be used. The locomotive main suspension comprised underhung leaf springs and in the first batch, Nos 4901–4980, spring compensating beams were fitted, but succeeding locomotives were built without, and the equipment was eventually removed from the earlier ones.

In addition to the smokebox saddle half-castings, two further saddles/frame stretchers were provided to carry the weight of the boiler. At the front end, this was positioned just ahead of the leading wheels, where the front frame extensions were bolted on, and provided support and mounting points for the valve rod rocking shafts. On the running boards at this point, sheet steel covers could be seen protecting the assembly. Inside the frames, a short link then connected with the forked end of the extension rod. Just behind the leading axlebox, on the inside faces of both frames, the link hanger was fixed, with the die block carried in the expansion link, which in its turn was connected to the reversing shaft via a link and arm. The reversing shaft itself was carried across the frames, with the right-hand end passing through the mild steel plate to the reversing shaft arm and ultimately the reach rod from the hand wheel in the cab. The eccentric sheaves were fixed to the leading coupled axle, with the fore gear and back gear eccentric rods attached to the top and bottom respectively of the expansion link.

The layout as shown in the diagram, was identical with other two-cylinder classes having 30in stroke. The only changes made were shorter or longer extension rods, valve rods, and rocking shaft drop arms. The valve rods were of only two different sizes, as applied to the GWR modern two-cylinder types, with the shorter variety fitted only to the 43XX 2-6-0s; longer style valve rods were used on the Saint, Hall, Modified Hall, Grange, Manor, and County classes.

A variation to Churchward's standardisation policies was the use of 6ft 0in coupled wheels with 20 spokes, a size not encountered in his original scheme of 1901. The wheels and axleboxes maintained the same standards under Collett, with cast-steel main axleboxes 10in long by 8¾in diameter, and having pressed-in white metal liners. The

very generous proportions of the GWR axleboxes was a key feature in their serviceability, and the later standard steam locomotives of British Railways, adopted similarly generous designs of plain bearing journal. The GWR method of balancing pairs of wheels on their axles, with molten lead poured into the space between plates rivetted onto the spokes, was an improvement on existing practice, and allowed greater accuracy in achieving the balance.

Simplicity was the keynote of much of Churchward's, and the Great Western's locomotive design principles, with plain, slab-sided coupling rods, and only the connecting rods fluted. Hall class locomotives had the standard brake rigging, with shoes behind the wheel treads on single hangers, and sanding was applied to the rear of the trailing coupled wheels, and in front of the leading wheels.

General Design Features

Above the footplate, the Halls were provided with the new Collett design side window cab, which was a rather more commodious affair than the old Churchward type. The layout of the controls on the footplate, for driver and fireman was almost identical with other two-cylinder classes, and grouped in front of the outermost line of the regulator handle. The hanging plate or valance, was joggled inwards just in front of the cab, in common with other types, and on the right-hand side supported and partly concealed the crosshead-driven air pump. On each side of the locomotive, a plate was fitted behind the slide bars to prevent any mud and grit that may have been thrown up by the bogie wheels from scoring the surfaces of the slide bars.

The coupled wheel splashers carried the decorative, half-round beading, with name-plates fixed over the driving wheel splashers. Later builds also carried a fire iron tunnel on the left-hand side alongside the firebox on the running boards, which footplate crews found useful for storing the longer fire irons.

At the front of the locomotive, where the smokebox joined the boiler barrel, a characteristic feature of GWR locomotives was the sheet metal cover over the lubrication pipes which emerged from beneath the boiler cladding at this point. Here the pipes, carrying oil to lubricate the cylinders,

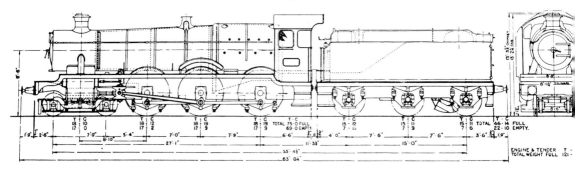

49XX Hall class 4-6-0

entered the smokebox through elbows, which could be easily disconnected if required during maintenance or repair. The system of cylinder lubrication, with the flow of oil controlled by the regulator was standardised around 1909. Originally, the feed pipes from the cab were carried under the boiler cladding on the right-hand side only, but in later years the feeds were divided further back, and the pipes were taken down both sides of the boiler. While this was a standard arrangement, the covers at the

No 5983 *Henley Hall*, in pristine condition after overhaul in 1959 at Swindon Works, awaiting steaming. On the footplate, the standard GWR sight-feed lubricator is just visible. *L. C. Jacks*

smokebox-boiler joint were not present on all locomotives.

In the early 1930s No 4952 was fitted with an automatic lubrication system, for use when the locomotive was drifting with steam shut off. The locomotive carried a large sheet metal cover over the apparatus, which was mounted on the right-hand side on the running boards next to the smokebox, but it had been removed by 1936. Mechanical lubricators instead of the standard sight feed arrangement were tried on Nos 4941 and 4950 from 1931/32, but no further examples were fitted until No 4905 received a mechanical lubricator in 1947.

Chimneys, and the combined safety valve and top feed mounted on the boiler, were perhaps the most distinctive feature of many GWR locomotives. In the Hall class, the short pattern safety valve bonnet had been chosen and was carried from new. Typical

copper-capped chimneys were about 1ft 10in in cast iron, and as standard for tender engines a small wind/smoke deflector or capuchon was fitted to the chimney rim. Isolated examples could be found in service without this small detail as a result of interchangeability of boilers.

Construction
The production series of Hall class engines

A very short lived experiment was electric lighting, here oil burner 3904 (originally No 4972), is pictured with headlights on the lamp brackets, and generator on the footplate. *Locomotive & General Railway Photographs.*

were built in 12 batches, between 1928 and 1943, and were covered by only three diagrams. Demonstrating what little variation there was between each build, the different diagrams show only such changes as tender design to be major alterations. Despite the

One-time oil burner No 5955 *Garth Hall* in later days, converted back to burn coal, and paired with a Hawksworth 4,000-gallon tender. *L. C. Jacks.*

Batch	Built	Lot Numbers	Running Numbers
1	1928–30	254	4901–4980
2	1931	268	4981–4999, 5900
3	1931	275	5901–5920
4	1933	281	5921–5940
5	1935	290	5941–5950
6	1935–36	297	5951–5965
7	1937	304	5966–5975
8	1938	311	5976–5985
9	1939–40	327	5986–5995
10	1940	333	5996–5999, 6900–6905
11	1940–41	338	6906–6915
12	1941–43	340	6916–6958

numerical strength of the class, there were surprisingly few variations, one being the words 'Hall Class' painted on the middle splashers of Nos 6916–6958, which were put to traffic nameless as a wartime economy – nameplates were provided later. From No 5922 onwards upper lamp brackets were fixed to the top of the smokebox door. During World War II, Nos 6906–6958 were built without side windows in the cab, although these were eventually inserted between 1945 and 1948.

GWR Audible Signalling apparatus was fitted to the class from No 4921 onwards as built, and retrospectively to the remainder of the class in 1930, including *Saint Martin*. Towards the end of GWR ownership, electric lighting equipment was fitted experimentally to No 4972 in 1947, and two years later, No 5904 was similarly modified.

The locomotive diagrams were: A1–4901 class with 3,500-gallon Churchward tender; A2–4901 class with 3,500-gallon Collett tender; A3–4901 class with 4,000-gallon tender.

Eighty-four Halls were included in the oil-burning locomotives conversion plan in 1946, but only eleven were converted, with the first, No 5955 *Garth Hall*, being modified in June 1946. The remaining ten locomotives were subsequently equipped as oil burners in April and May 1947, just months before nationalisation, and were renumbered in a 39XX series. The conversion of the Halls followed an earlier, successful experiment with 18 of the 28XX class. The average life of the Halls as oil burners was only two years, with all being re-converted to coal burning by April 1950. Those converted were:

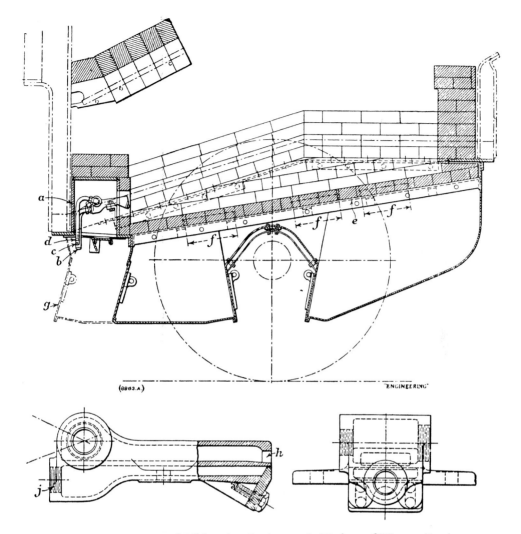

Arrangement of Oil-burning Equipment in Firebox of Western Region Locomotive.

a Burner Box.
b Fuel-oil Pipe.
c Steam Pipe.
d Steam pipe for cleaning burner.
e Firebox Floor Plate.

f Air openings.
g Ashpan before conversion to oil burning.
h Oil orifice.
j Steam pipe connection to burner.

Running No	New No	Date converted	Date reconverted
4968	3900	5/1947	3/1949
4971	3901	5/1947	4/1949
4948	3902	5/1947	9/1948
4907	3903	5/1947	4/1950
4972	3904	5/1947	10/1948
5955	3950	6/1946	10/1948
5976	3951	4/1947	11/1948
6957	3952	4/1947	3/1950
6953	3953	4/1947	9/1948
5986	3954	5/1947	2/1950
6949	3955	5/1947	4/1949

On the converted locomotives, the firebox was modified by the removal of the grate and by fitting a steel plate floor lined with firebricks, with air spaces from the former ashpan, which retained the original damper gear. On the footplate, the controls included a steam manifold with control valves, and a fuel valve for regulating the oil supply to the single burner, mounted at the bottom front end of the firebox. The controls were on the fireman's side, and regulated steam supplies to the burner, burner cleaner, smokebox blower, locomotive and tender oil heaters.

On the tender, a 1,950-gallon fuel tank was installed, and provided with heater coils to maintain the viscosity of the oil at a suitable level. On top of the tank, a filler hole and depth indicator were fitted, with a thermometer and steam heater manifold at the front. Fuel from the tank flowed through a 2in shut-off cock and strainer and via a flexible connection to the locomotive, and finally through a 1½in bore pipe to the burner at the front of the firebox.

The burner was carried at foundation ring level in a small chamber with a bottom damper door which could be pre-adjusted to a suitable opening. In the burner the upper opening, measuring 2in × ½in, allowed the oil to pass through at a temperature of around 180°F, while immediately below this was the much smaller opening, only 2½in × 0.018in, for the atomising steam jet. As the oil ran out of the upper opening, it came into contact with the high-pressure steam jet, and was turned into a fine spray (atomised).

The oil burning members of the class were almost 13¾ tons heavier than their coal-fired counterparts, the tender accounting for an additional 6 tons 9 cwt. The weight distribution was affected too, resulting in the following changes to axle loading; bogie, 18

The first of the oil-fired locomotives, No 5955 Garth Hall, with shutters over the cabside windows, and, clearly visible, the large oil tank carried in the tender coal space. Locomotive & General Railway Photographs.

tons 10 cwt; coupled wheels, 18 tons 15 cwt, 19 tons 3 cwt, 19 tons 3 cwt – making an all-up weight of 75 tons 11 cwt for the locomotive alone. In the right hands they were superior to their coal fired equivalents, and in the West of England some of their finest work was performed over the hilly routes that abound. While they were still classed as 'red' route, power class D, like the normal Halls, Swindon issued diagram A21 to cover the 39XX series of oil burners.

Operations

By the end of 1947 Hall class locomotives were allocated to more than 30 GWR depots. Initially the first 14 Halls were sent to the West Country, allocated to Penzance and Laira (Plymouth), but as new locomotives appeared, they very rapidly spread throughout the system. Workings normally associated with Halls were as varied as the names they carried – freight, empty stock and stopping passenger, to the most prestigious express services. In the far west, only the Cornish Riviera Express was without a Hall class locomotive at its head in the early days, but in later years even this omission was overcome.

During World War II, some of those in especially good mechanical condition were permitted to haul trains which exceeded the normal loading for the class. Locomotives with this allowance had a white X suffix to the running number. During the hostilities

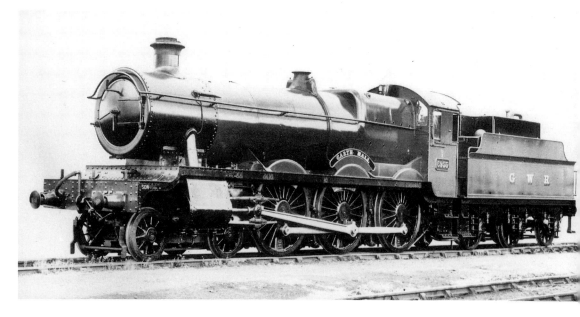

No 4911 *Bowden Hall* received a direct hit from a bomb at Keyham; the remains were taken to Swindon and condemned in June 1941. The Hall class later passed into British Railways ownership with only this gap in their numbers.

Although in BR days steam traction was superseded by diesel earlier than elsewhere, the Hall class remained largely intact in the first few years until about 1961. Between 1948 and 1961 only 12 were scrapped. No 4946 *Moseley Hall* was the first locomotive to carry any indication of its new ownership, when in January 1948 it was turned out from Swindon in GWR livery, but with the tender lettered BRITISH RAILWAYS in full. Nos 4904/46/51 and 5958 were given the temporary W suffix to the running numbers, painted on the cabside below the number plate. No 6910 was the first to receive the lined black livery adopted by the new organisation in those early years for the whole class. During the 1950s, a much greater number of Halls were at work on main line passenger services than had been seen previously. Rapid dieselisation in the 1960s quickly reduced their ranks, with only 50 survivors in the 69XX series in 1965. This was their final year of service, with the last stabled at Banbury, Bristol, Didcot, Cardiff, Gloucester, Oxford, Severn Tunnel Junction, Worcester and Tyseley.

Hall class 4-6-0 – Building and withdrawal dates

Running No	Name	Built	Withdrawn	Scrapping/disposal
4900	Saint Martin	12/24*	4/59	Swindon Works
4901	Adderley Hall	12/28	9/60	Swindon Works
4902	Aldenham Hall	12/28	9/63	Cohen, Swansea
4903	Astley Hall	12/28	10/64	Steel Supply Co, Swansea
4904	Binnegar Hall	12/28	12/63	Birds, Risca
4905	Barton Hall	12/28	11/63	Cashmore, Great Bridge
4906	Bradfield Hall	1/29	9/62	Cashmore, Great Bridge
4907	Broughton Hall	1/29	8/63	King, Norwich
4908	Broome Hall	1/29	10/63	Cashmore, Newport
4909	Blakesley Hall	1/29	9/62	Cashmore, Newport
4910	Blaisdon Hall	1/29	12/63	King, Norwich
4911	Bowden Hall	2/29	6/41**	Swindon Works
4912	Berrington Hall	2/29	8/62	Swindon Works
4913	Baglan Hall	2/29	9/62	Hayes, Bridgend
4914	Cranmore Hall	2/29	12/63	Birds, Risca
4915	Condover Hall	2/29	2/63	Swindon Works
4916	Crumlin Hall	2/29	8/64	Cohen, Swansea
4917	Crosswood Hall	3/29	9/62	Cashmore, Newport
4918	Dartington Hall	3/29	6/63	Hayes, Bridgend
4919	Donnington Hall	3/29	10/64	Steel Supply Co, Swansea
4920	Dumbleton Hall	3/29	12/65	Preserved
4921	Eaton Hall	4/29	9/62	King, Norwich
4922	Enville Hall	4/29	7/63	Cashmore, Newport
4923	Evenley Hall	4/29	5/63	Cashmore, Great Bridge
4924	Eydon Hall	5/29	10/63	Coopers Metals, Sharpness
4925	Eynsham Hall	5/29	8/62	Swindon Works
4926	Fairleigh Hall	5/29	9/61	Swindon Works
4927	Farnborough Hall	5/29	9/63	Cohen, Swansea
4928	Gatacre Hall	5/29	12/63	Cashmore, Newport
4929	Goytrey Hall	5/29	3/65	Cashmore, Newport
4930	Hagley Hall	5/29	12/63	Preserved
4931	Hanbury Hall	5/29	7/62	Hayes, Bridgend
4932	Hatherton Hall	6/29	11/64	Hayes, Bridgend
4933	Himley Hall	6/29	8/64	Cashmore, Great Bridge
4934	Hindlip Hall	6/29	9/62	Cashmore, Newport
4935	Ketley Hall	6/29	3/63	Cashmore, Newport
4936	Kinlet Hall	6/29	1/64	Preserved
4937	Lanelay Hall	6/29	12/62	Swindon Works
4938	Liddington Hall	6/29	2/63	Swindon Works
4939	Littleton Hall	7/29	2/63	Wolverhampton Works
4940	Ludford Hall	7/29	11/59	Swindon Works
4941	Llangedwyn Hall	7/29	10/62	Cashmore, Newport
4942	Maindy Hall	7/29	12/63	Preserved
4943	Marrington Hall	7/29	12/63	Birds, Swansea
4944	Middleton Hall	7/29	9/62	King, Norwich
4945	Milligan Hall	8/29	11/61	Swindon Works

Running No	Name	Built	Withdrawn	Scrapping/disposal
4946	Moseley Hall	8/29	6/63	Cohen, Swansea
4947	Nanhoran Hall	8/29	9/62	King, Norwich
4948	Northwick Hall	8/29	9/62	Cashmore, Newport
4949	Packwood Hall	8/29	9/64	Birds, Llanelly
4950	Patshull Hall	8/29	5/64	Central Wagon Co, Swansea
4951	Pendeford Hall	7/29	6/64	Cashmore, Newport
4952	Peplow Hall	8/29	9/62	Hayes, Bridgend
4953	Pitchford Hall	8/29	4/63	Preserved
4954	Plaish Hall	8/29	11/64	Cashmore, Great Bridge
4955	Plaspower Hall	8/29	10/63	Cohen, Swansea
4956	Plowden Hall	9/29	7/63	Cashmore, Newport
4957	Postlip Hall	9/29	3/62	Swindon Works
4958	Priory Hall	9/29	9/64	Cashmore, Newport
4959	Purley Hall	9/29	12/64	Buttigieg, Newport
4960	Pyle Hall	9/29	9/62	Cashmore, Newport
4961	Pyrland Hall	11/29	12/62	Swindon Works,
4962	Ragley Hall	11/29	10/65	Birds, Risca
4963	Rignall Hall	11/29	6/62	Swindon Works
4964	Rodwell Hall	11/29	10/63	Cashmore, Newport
4965	Rood Ashton Hall	11/29	3/62	Swindon Works
4966	Shakenhurst Hall	11/29	11/63	Cashmore, Great Bridge
4967	Shirenewton Hall	12/29	9/62	Hayes, Bridgend
4968	Shotton Hall	12/29	7/62	Cashmore, Newport
4969	Shrugborough Hall	12/29	9/62	Swindon Works
4970	Sketty Hall	12/29	7/63	Cohen, Swansea
4971	Stanway Hall	1/30	8/62	Hayes, Bridgend
4972	Saint Brides Hall	1/30	2/64	Hayes, Bridgend
4973	Sweeney Hall	1/30	7/62	Hayes, Bridgend
4974	Talgarth Hall	1/30	4/62	Swindon Works
4975	Umberslade Hall	1/30	9/63	Cashmore, Newport
4976	Warfield Hall	1/30	5/64	Central Wagon Co, Wigan
4977	Watcombe Hall	1/30	5/62	Swindon Works
4978	Westwood Hall	2/30	9/64	Hayes, Bridgend
4979	Wootton Hall	2/30	12/63	Sold privately
4980	Wrottesley Hall	2/30	7/63	Cashmore, Newport
4981	Abberley Hall	12/30	10/63	Cashmore, Newport
4982	Acton Hall	1/31	5/62	Swindon Works
4983	Albert Hall	1/31	12/63	Preserved
4984	Albrighton Hall	1/31	9/62	Hayes, Bridgend
4985	Allesley Hall	1/31	9/64	Cashmore, Newport
4986	Aston Hall	1/31	5/62	Swindon Works
4987	Brockley Hall	1/31	4/62	Swindon Works
4988	Bulwell Hall	1/31	2/64	Swindon Works
4989	Cherwell Hall	2/31	11/64	Hayes, Bridgend
4990	Clifton Hall	2/31	4/62	Swindon Works
4991	Cobham Hall	2/31	12/63	Cashmore, Great Bridge
4992	Crosby Hall	2/31	4/65	Birds, Risca
4993	Dalton Hall	2/31	2/65	Hayes, Bridgend
4994	Downton Hall	2/31	3/63	Cashmore, Newport
4995	Easton Hall	2/31	6/62	Swindon Works
4996	Eden Hall	3/31	9/63	Coopers Metals, Sharpness
4997	Elton Hall	3/31	10/61	Swindon Works
4998	Eyton Hall	3/31	10/63	Cohen, Swansea
4999	Gopsal Hall	3/31	9/62	Cashmore, Newport
5900	Hinderton Hall	3/31	12/63	Preserved
5901	Hazel Hall	5/31	6/64	Cashmore, Great Bridge
5902	Howick Hall	5/31	12/62	Swindon Works
5903	Keele Hall	5/31	9/63	Cashmore, Newport
5904	Kelham Hall	5/31	11/63	Hayes, Bridgend
5905	Knowsley Hall	5/31	7/63	Coopers Metals, Sharpness
5906	Lawton Hall	5/31	5/62	Swindon Works
5907	Marble Hall	5/31	11/61	Swindon Works
5908	Moreton Hall	6/31	7/63	Cashmore, Newport
5909	Newton Hall	6/31	7/62	Hayes, Bridgend
5910	Park Hall	6/31	9/62	Cashmore, Great Bridge
5911	Preston Hall	6/31	9/62	Hayes, Bridgend
5912	Queen's Hall	6/31	12/62	Swindon Works
5913	Rushton Hall	6/31	5/62	Cashmore, Newport
5914	Ripon Hall	7/31	1/64	Cashmore, Newpot
5915	Trentham Hall	7/31	1/60	Swindon Works
5916	Trinity Hall	7/31	7/62	Cashmore, Great Bridge
5917	Westminster Hall	7/31	9/62	King, Norwich
5918	Walton Hall	7/31	9/62	King, Norwich
5919	Worsley Hall	7/31	8/63	Cashmore, Newport
5920	Wycliffe Hall	8/31	1/62	Swindon Works
5921	Bingley Hall	5/33	1/62	Swindon Works
5922	Caxton Hall	5/33	1/64	Swindon Works

Running No	Name	Built	Withdrawn	Scrapping/disposal
5923	Colston Hall	5/33	12/63	Cox & Danks, North Acton
5924	Dinton Hall	5/33	12/63	Birds, Risca
5925	Eastcote Hall	5/33	10/62	Swindon Works
5926	Grotrian Hall	6/33	9/62	Cashmore, Great Bridge
5927	Guild Hall	6/33	10/64	Cashmore, Great Bridge
5928	Haddon Hall	6/33	5/62	Swindon Works
5929	Hanham Hall	6/33	10/63	Hayes, Bridgend
5930	Hannington Hall	6/33	9/62	Cashmore, Great Bridge
5931	Hatherley Hall	6/33	9/62	Swindon Works
5932	Haydon Hall	6/33	10/65	Cashmore, Newport
5933	Kingsway Hall	6/33	8/65	Birds, Bynea
5934	Kneller Hall	6/33	5/64	Hayes, Bridgend
5935	Norton Hall	7/33	1/62	Coopers Metals, Sharpness
5936	Oakley Hall	7/33	1/65	Swindon Works
5937	Stanford Hall	7/33	11/63	Cashmore, Newport
5938	Stanley Hall	7/33	5/63	Wolverhampton Works
5939	Tangley Hall	7/33	10/64	Hayes, Bridgend
5940	Whitbourne Hall	8/33	9/62	Cashmore, Newpot
5941	Campion Hall	2/35	7/62	Swindon Works
5942	Doldowlod Hall	2/35	12/63	Cashmore, Great Bridge
5943	Elmdon Hall	3/35	6/63	Coopers Metals, Sharpness
5944	Ickenham Hall	3/35	4/63	Cashmore, Newport
5945	Leckhampton Hall	3/35	4/63	Swindon Works
5946	Marwell Hall	3/35	7/62	Hayes, Bridgend
5947	Saint Benet's Hall	3/35	7/62	Swindon Works
5948	Siddington Hall	3/35	8/63	Hayes, Bridgend
5949	Trematon Hall	4/35	5/61	Swindon Works
5950	Wardley Hall	12/35	11/61	Swindon Works
5951	Clyffe Hall	12/35	4/64	Cashmore, Great Bridge
5952	Cogan Hall	12/35	6/64	Sold to GWS Locomotive Group
5953	Dunley Hall	12/35	10/62	Swindon Works
5954	Faendre Hall	12/35	10/63	Coopers Metals, Sharpness
5955	Garth Hall	12/35	4/65	Birds, Risca
5956	Horsley Hall	12/35	3/63	Swindon Works
5957	Hutton Hall	12/35	7/64	Birds, Swansea
5958	Knolton Hall	1/36	3/64	Hayes, Bridgend
5959	Mawley Hall	1/36	9/62	Cashmore, Great Bridge
5960	Saint Edmund Hall	1/36	9/62	King, Norwich
5961	Toynbee Hall	6/36	8/65	Birds, Bridgend
5962	Wantage Hall	7/36	11/64	Cashmore, Great Bridge
5963	Wimpole Hall	7/36	6/64	Birds, Risca
5964	Wolseley Hall	7/36	9/62	Swindon Works
5965	Woollas Hall	8/36	7/62	Cashmore, Great Bridge
5966	Ashford Hall	3/37	9/62	King, Norwich
5967	Bickmarsh Hall	3/37	6/64	Preserved
5968	Corey Hall	3/37	9/62	Cashmore, Newport
5969	Honington Hall	4/37	8/62	Swindon Works
5970	Hengrave Hall	4/37	11/63	Cashmore, Newport
5971	Merevale Hall	4/37	12/65	Cashmore, Newport
5972	Olton Hall	4/37	12/63	Preserved
5973	Rolleston Hall	5/37	9/62	Swindon Works
5974	Wallsworth Hall	4/37	12/64	Swindon Works
5975	Winslow Hall	5/37	7/64	Cashmore, Newport
5976	Ashwicke Hall	9/38	7/64	Cashmore, Newport
5977	Beckford Hall	9/38	8/63	Cashmore, Newport
5978	Bodinnick Hall	9/38	10/63	Coopers Metals, Sharpness
5979	Cruckton Hall	9/38	11/64	Hayes, Bridgend
5980	Dingley Hall	9/38	9/62	Cashmore, Newport
5981	Frensham Hall	10/38	9/62	Hayes, Bridgend
5982	Harrington Hall	10/38	9/62	King, Norwich
5983	Henley Hall	10/38	4/65	Birds, Bridgend
5984	Linden Hall	10/38	1/65	Buttigieg, Newport
5985	Mostyn Hall	10/38	9/63	Cashmore, Newport
5986	Arbury Hall	11/39	9/63	Cashmore, Newport
5987	Brocket Hall	11/39	1/64	Swindon Works
5988	Bostock Hall	11/39	10/65	Cashmore, Great Bridge
5989	Cransley Hall	12/39	7/62	Hayes, Bridgend
5990	Dorford Hall	12/39	1/65	Friswell, at Banbury shed
5991	Gresham Hall	12/39	7/64	Cashmore, Great Bridge
5992	Horton Hall	12/39	8/65	Birds, Swansea
5993	Kirby Hall	12/39	5/63	Swindon Works
5994	Roydon Hall	12/39	3/63	Cashmore, Newport
5995	Wick Hall	1/40	4/63	Swindon Works
5996	Mytton Hall	6/40	8/62	Swindon Works
5997	Sparkford Hall	6/40	7/62	Swindon Works
5998	Trevor Hall	6/40	3/64	Cashmore, Great Bridge
5999	Wollaton Hall	6/40	9/62	Swindon Works

Running No	Name	Built	Withdrawn	Scrapping/disposal
6900	Abney Hall	6/40	10/64	Hayes, Bridgend
6901	Arley Hall	7/40	6/64	Birds, Risca
6902	Butlers Hall	7/40	5/61	Swindon Works
6903	Belmont Hall	7/40	9/65	Birds, Long Marston
6904	Charfield Hall	7/40	2/65	Friswell, at Banbury shed
6905	Claughton Hall	7/40	6/64	Hayes, Bridgend
6906	Chicheley Hall	11/40	4/65	Cashmore, Great Bridge
6907	Davenham Hall	11/40	2/65	Cashmore, Great Bridge
6908	Downham Hall	11/40	7/65	Cashmore, Newport
6909	Frewin Hall	11/40	6/64	Cashmore, Newport
6910	Gossington Hall	12/40	10/65	Birds, Risca
6911	Holker Hall	1/41	4/65	Birds, Long Marston
6912	Helmster Hall	1/41	2/64	Hayes, Bridgend
6913	Levens Hall	2/41	6/64	Birds, Swansea
6914	Langton Hall	2/41	4/64	Birds, Risca
6915	Mursley Hall	2/41	2/65	Friswell, at Banbury shed
6916	Misterton Hall	6/41	8/65	Friswell, at Banbury shed
6917	Oldlands Hall	6/41	9/65	Friswell, at Banbury shed
6918	Sandon Hall	6/41	9/65	Birds, Long Marston
6919	Tylney Hall	6/41	8/63	Swindon Works
6920	Barningham Hall	7/41	12/63	Birds, Swansea
6921	Borwick Hall	7/41	10/65	Friswell, at Banbury shed
6922	Burton Hall	7/41	4/65	Birds, Long Marston
6923	Croxteth Hall	7/41	12/65	Cashmore, Newport
6924	Grantley Hall	8/41	10/65	Friswell, at Banbury shed
6925	Hackness Hall	8/41	11/64	Cashmore, Great Bridge
6926	Holkham Hall	11/41	6/65	Cashmore, Great Bridge
6927	Lilford Hall	11/41	10/65	Birds, Risca
6928	Underley Hall	11/41	6/65	Cohen, Kettering
6929	Whorlton Hall	11/41	10/63	Cohen, Swansea
6930	Aldersley Hall	11/41	10/65	Friswell, at Banbury shed
6931	Aldborough Hall	12/41	10/65	Birds, Risca
6932	Burwarton Hall	12/41	12/65	Cashmore, Newport
6933	Birtles Hall	12/41	11/64	Cashmore, Great Bridge
6934	Beachamwell Hall	12/41	10/65	Cashmore, Great Bridge
6935	Browsholme Hall	12/41	2/65	Swindon Works
6936	Breccles Hall	7/42	11/64	Hayes, Bridgend
6937	Conyngham Hall	7/42	12/65	Cashmore, Newport
6938	Corndean Hall	7/42	3/65	Hayes, Bridgend
6939	Calveley Hall	7/42	10/63	Hayes, Bridgend
6940	Didlington Hall	8/42	5/64	Cashmore, Great Bridge
6941	Fillongley Hall	8/42	4/64	Cohen, Swansea
6942	Eshton Hall	8/42	12/64	Cashmore, Newport
6943	Farnley Hall	8/42	12/63	Birds, Risca
6944	Fledborough Hall	9/42	12/65	Buttigieg, Newport
6945	Glasfryn Hall	9/42	9/64	Woodfield, Cadoxton
6946	Heatherden Hall	12/42	6/64	Birds, Risca
6947	Helmingham Hall	12/42	10/65	Cashmore, Newport
6948	Holbrooke Hall	12/42	12/63	Cashmore, Great Bridge
6949	Haberfield Hall	12/42	5/61	Swindon Works
6950	Kingsthorpe Hall	12/42	6/64	Cashmore, Newport
6951	Impney Hall	2/43	12/65	Cashmore, Great Bridge
6952	Kimberley Hall	2/43	12/65	Cashmore, Great Bridge
6953	Leighton Hall	2/43	12/65	Cashmore, Newport
6954	Lotherton Hall	3/43	5/64	Cohen, Swansea
6955	Lydcott Hall	3/43	2/65	Swindon Works
6956	Mottram Hall	3/43	12/65	Cashmore, Newpot
6957	Norcliffe Hall	4/43	10/65	Friswell, at Banbury shed
6958	Oxburgh Hall	4/43	6/65	Cohen, Swansea

*   No 4900 – 12/24 was date of rebuilding of No 2925 *Saint Martin* as prototype for Hall class.
** No 4911 – withdrawn after damage by enemy action in World War II.

CHAPTER FIVE

COLLETT'S LATER DESIGNS – THE GRANGE AND MANOR CLASSES

C. B. Collett succeeded Churchward as Great Western Railway Chief Mechanical Engineer in 1922, and presided over the introduction of no fewer than 20 new locomotive classes. Only two of these came into the category of two-cylinder tender locomotives, both of which were introduced as successors to the 43XX 2-6-0s. The Grange class 4-6-0 was introduced in 1936, as power class D and restricted to 'red' routes, although the design's ancestry can be traced back to a 5ft 8in 4-6-0 outlined by Churchward in his master plan of 1901. The Manor class emerged from Swindon in 1938 as a lighter derivative, also in power class D, but with a wider availability on 'blue' routes. Only 20 Manors were built by the GWR, with a further 10 appearing after nationalisation in 1950. Both classes were ultimately successful, the Granges immediately, but the smaller Manor class was plagued with steaming difficulties, and began life with a bad reputation in this respect. The Granges, with the Standard No 1 boiler were still very much a Churchward type, while the Manors

introduced an entirely new design of boiler, though still on Churchward principles. In the early 1950s, some of these principles were not found to hold so good for a number of types, and re-draughting improved their performance substantially.

Grange Class

These locomotives – a total of 80 was built in all – had been advocated as the mixed traffic type required by the operating department, before the eminently successful Hall class was built. Making use of certain components from redundant 43XX 2-6-0s – wheels and motion in particular – the Grange class locomotives were completed at a cost of around £3,900 each. Officially 'new' the locomotives may have been, but almost the entire class was fitted with reconditioned 3,500-gallon capacity tenders; Nos 6801 and 6835 for instance were paired with tenders

The operating department's mixed traffic type, with the first of Collett's new design, No 6800 *Arlington Grange*, seen here with its secondhand tender. *L. C. Jacks*

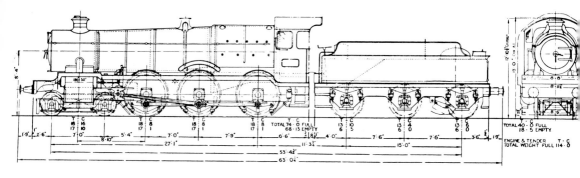

68XX Grange class 4-6-0

built in 1907 and 1911 respectively. Having introduced a new class of locomotive, Collett had provided an economic compromise, with the use of reconditioned wheels and motion, and second-hand tenders. The Grange was the answer to the operating department's request for a 43XX 2-6-0 with a Standard No 1 boiler; but they were not simply a smaller wheeled variant of the Hall class.

Boiler details

The boiler and firebox assembly of the Grange class locomotives was again the Standard No 1 boiler, similar in general features with those fitted to the Hall class. The 14ft 10in barrel, built from two tapered rings, housed the 176 tubes, and 84 superheater elements in two rows of 5⅛in diameter flues. No fewer than 629 of this type of boiler were constructed from 1922 onwards. Among others, these boilers were interchangeable

between the Saint, Hall, and Grange classes, and boiler changes did from time to time result in the appearance of Hall type chimneys on Grange class locomotives. Fireboxes too were the same as on the Halls, with the familiar Churchward design of Belpaire form – widened water legs to improve circulation, the sides and top tapered inwards towards the cab, improving forward visibility.

Short standard pattern safety valve bonnet and top feed combination was positioned in the centre of the rear tapered ring, with the valves set to lift at 225 lb/sq in pressure. Feed water entered the boiler and was deflected by means of lightweight plates carried beneath the valves inside the boiler, and reduced the tendency of pitting on the boiler plates to a minimum. The feed water came into contact with the boiler water at a temperature estimated to be 40°F below that of the boiler water.

The standard Swindon two row No 3 superheater was fitted in the Churchward

Grange class 4-6-0 – Leading details

Wheel arrangement	4–6–0			Heating surface		
Running numbers	6800–6879			– tubes	1686.60 sq ft	
Lot No	308			– firebox	154.78 sq ft	
Built	8/1936–5/1939			– total	1841.38 sq ft	
Withdrawn	10/1960–12/1965			– superheater	262.62 sq ft	

Wheel diameter				Firebox – length o/s	9ft 0in	
– coupled	5ft 8in			– width	5ft 9in/4ft 0in	
– bogie	3ft 0in					
Wheelbase	27ft 1in			Grate area	27.07 sq ft	

Axle load	Tons Cwt leading	Tons Cwt driving	Tons Cwt trailing	Boiler pressure	225 lb/sq in
– coupled	18 08	18 08	18 08		
	Tons Cwt			Tractive effort	28,875 lb
– bogie	18 16				

Boiler			Fuel capacities – coal		6 tons		
– diameter	4ft 10⅞in tapering to 5ft 6in		– water		3,500 gallons		
– length	14ft 10in						
– tubes, small	176 × 2in o/d	Weights	Full		Empty		
– tubes, large	14 × 5⅛in o/d		Tons	Cwt	Tons	Cwt	
– superheater elements	84 × 1in o/d	– locomotive	74	00	68	13	
		– tender	40	00	18	05	
Cylinders – bore × stroke	18½in × 30in	– total	114	00	86	18	

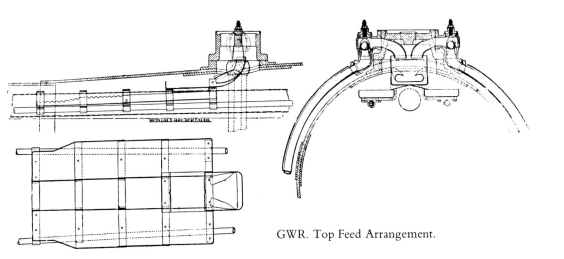

GWR. Top Feed Arrangement.

boiler designs, as a result of practical experimenting. Although criticised in principle for offering only a low degree of superheat, the method which was sufficient to remove the tendency to condensation in the cylinders was believed adequate for current and prospective applications. Among the reasons advanced for its use were that it should be available for all classes of locomotive, even the older types, fitted with bronze slide valves. For maintenance, each group of elements could be withdrawn without disturbing others, and was the most practical arrangement in the two row layout. Churchward's view was that superheating should be provided only at a sufficient temperature to prevent condensation, and avoid lubrication

troubles. The temperatures reached in this practice – typically about 120°F to 180°F – resulted in a final steam temperature of around 500°F. All the Grange class were constructed with this Swindon standard arrangement.

At the front end, the smokebox had reached its standard length (6ft 0in), and was again very similar to the Halls. In original form, the blastpipe incorporated the jumper ring, whose action was intended to lessen the effects of the very sharp blast from Great Western two-cylinder locomotives when operating at long cut-off. The annular holes in the blast nozzle casting provided an outlet for the blower and ejector, exhausting directly into the chimney.

Frames, Wheels and Motion
The main frames of the Grange class were manufactured in the traditional Swindon manner from mild steel plate, terminating just behind the cylinders, with the front extension frames bolted in place. The front stiffening struts applied to the 43XX class were not used on the Granges; there were also differences in the construction and fitting of cylinder castings. The arrangement of the latter allowed straight extension frames to be used, in comparison with the 43XX which had curved extensions which provided less strength at this point, requiring the provision of stiffening struts.

Wheelsets for the Granges came from a 'reconditioned' pool, obtained from withdrawn 2-6-0s Nos 4300/1/4/5/6/8–11/13–15/17/23/24/27–34/36/38–41/44–48/50/51. It has been suggested that the wheels and

Blast Nozzle fitted
with Jumper Ring.

A. Movable Jumper Ring.
B. Fixed Blast Orifice.
C. Fixed Blast Pipe Cap.

motion from individual 43XX 2-6-0s were used for individual Granges, but this seems unlikely since Swindon already had 12 sets of wheels and motion by the time the first Grange was built. From the 40 withdrawn 43XX 2-6-0s in 1936 Swindon had enough parts to build Grange class Nos 6800–6829 between August 1936 and March 1937. The wheel diameter of 5ft 8in was a Churchward standard laid down in 1901, with the traditional cast-steel axleboxes of generous proportions, with pressed in white metal inserts. Hornblock, and wheel hub bearing faces were fitted with renewable bronze liners. The 18-spoke coupled wheels had individual balance weights of lead, sandwiched between pairs of steel plates, with the individual wheelsets being balanced on Swindon's balancing machine, before being fitted to the locomotive.

The 18½in by 30in cylinders were constructed and fitted along standard lines, but

An attractive view of No 6853 *Morehampton Grange*, stopped at signals at Fenny Compton on a Sunday morning in 1953. Raised footplating over the cylinders was necessary for correct alignment, with the larger steamchests used on these locomotives. *L. C. Jacks*

whereas in previous classes the steamchests partially surrounded the cylinders proper, a change was made in the new Grange class. Most notable difference was the larger steam chest, offset above the valve, which accounted for the raised footplating at this point. The steamchests were separate from the cylinders, and some of the first Granges, like the Halls, were at first fitted with 9in diameter inside admission piston valves. Later, the whole class was fitted with the standard 10in diameter valve. In order to use the same valve motion as the 43XX, Grange class cylinders were set 2½in lower in the frames, in order to make the cylinder and driving wheel centre lines coincide, with the valve centres at the same height above rail level as in the moguls. A consequence of this was the greater length of the inlet and exhaust ports between cylinder and valve. Clearance volume – the difference between total and piston swept volume – was greater than in the Hall class, and was a contributing factor in the new 4-6-0s reputation for smooth running, although it did not lead to any significant increase in steam consumption.

Supporting a front end weight of 18 tons 16 cwt was the standard DeGlehn–Churchward design of bar-frame bogie, with 3ft 0in wheels and the primary suspension of inverted leaf springs transmitting the load by way of ball-and-socket assemblies at either side. Side control was effected through coil springs. Steel plates were fixed behind the slidebars, preventing mud and grit being thrown up onto the surface of the bars by the bogie wheels, while on the right-hand side the vacuum pump was attached to the valance, driven by the crosshead.

Detail features

The locomotive main suspension was by means of underhung leaf spring, with equalising gear, and single-hanger brake shoes carried behind each coupled wheel. Sanding was arranged to deliver in front of the leading wheels and behind the trailing pair. It is an interesting point that, in the diagram issued for new Grange class locomotives, the three coupled wheelsets supported identical axle loads of 18 tons 8 cwt in working order.

Above the frames, platework provided smaller wheel splashers, with nameplates over the driving wheel, and on the fireman's side a fire iron tunnel was provided from the outset. The first four Granges carried cast-iron chimneys, but later builds sported the smaller typical Grange chimney with copper cap. Also at the front end, upper lamp brackets were fixed to the top of the smokebox door, from new. AWS gear was another item of equipment carried by Granges from new, as was the whistle shroud, on top of the fire-box immediately in front of the cab spectacle plate.

The cab was a typical Collett design incorporating side windows, and a rear handrail stanchion extending from floor to roof, which also acted as additional support for the latter. GWR cabs in general had a fairly high arc roof profile, in comparison with many other companies, and the footplate layout of the Grange class conformed to well-established Swindon practice.

Construction

The 80 locomotives of this new class were built to Swindon Lot No 308 between 1936 and 1939. The first 30 were delivered at a steady rate between August 1936 and March 1937; a five-month gap then followed before deliveries recommenced in August to December of that year. Between No 6859's appearance in December 1937 and No 6860 in February 1939 there was a gap of more than a year, and the final 20 were built between February and May 1939. Construction was to have continued, along with the new Manor class, until gradually all the 43XX moguls had been replaced. Unfortunately World War II interrupted this programme which was never restarted, although a further batch of the smaller Manor class 4-6-0s was built in 1950.

Grange class 4-6-0 – building and withdrawal dates

Running No	Name	Built	Withdrawn	Where scrapped
6800	*Arlington Grange*	8/1936	6/1964	Birds, Risca
6801	*Aylburton Grange*	8/1936	10/1960	Swindon Works
6802	*Bampton Grange*	9/1936	8/1961	Swindon Works
6803	*Bucklebury Grange*	9/1936	9/1965	Birds, Long Marston
6804	*Brockington Grange*	9/1936	8/1964	Birds, Risca
6805	*Broughton Grange*	9/1936	3/1961	Swindon Works
6806	*Blackwell Grange*	9/1936	10/1964	Swindon Works
6807	*Birchwood Grange*	9/1936	12/1963	King, Norwich
6808	*Beenham Grange*	9/1936	8/1964	Cohen, Kettering
6809	*Burghclere Grange*	9/1936	7/1963	Swindon Works
6810	*Blakemere Grange*	11/1936	10/1964	Birds, Bridgend
6811	*Cranbourne Grange*	11/1936	7/1964	Cashmore, Newport
6812	*Chesford Grange*	11/1936	2/1965	Swindon Works
6813	*Eastbury Grange*	12/1936	9/1965	Birds, Long Marston
6814	*Enborne Grange*	12/1936	12/1963	Swindon Works
6815	*Frilford Grange*	12/1936	11/1965	Buttigieg, Newport
6816	*Frankton Grange*	12/1936	7/1965	Birds, Long Marston
6817	*Gwenddwr Grange*	12/1936	4/1965	Birds, Risca
6818	*Hardwick Grange*	12/1936	4/1964	Birds, Risca
6819	*Highnam Grange*	12/1936	11/1965	Cashmore, Newport

Running No	Name	Built	Withdrawn	Where scrapped
6820	*Kingstone Grange*	1/1937	7/1965	Birds, Morriston
6821	*Leaton Grange*	1/1937	11/1964	Birds, Bridgend
6822	*Manton Grange*	1/1937	9/1964	Birds, Morriston
6823	*Oakley Grange*	1/1937	6/1965	Birds, Long Marston
6824	*Ashley Grange*	1/1937	4/1964	Swindon Works
6825	*Llanvair Grange*	2/1937	6/1964	Birds, Risca
6826	*Nannerth Grange*	2/1937	5/1965	Birds, Morriston
6827	*Llanfrechfa Grange*	2/1937	9/1965	Cashmore, Newport
6828	*Trellech Grange*	2/1937	7/1963	Swindon Works
6829	*Burmington Grange*	3/1937	11/1965	Cashmore, Newport
6830	*Buckenhill Grange*	8/1937	10/1965	Cohen, Kingsbury
6831	*Bearley Grange*	8/1937	10/1965	Cohen, Kettering
6832	*Brockton Grange*	8/1937	1/1964	Cohen, Morriston
6833	*Calcot Grange*	8/1937	10/1965	Cohen, Kettering
6834	*Dummer Grange*	8/1937	6/1964	Cashmore, Great Bridge
6835	*Eastham Grange*	9/1937	5/1963	Swindon Works
6836	*Estevarney Grange*	9/1937	8/1965	Birds, Bridgend
6837	*Forthampton Grange*	9/1937	7/1965	Birds, Morriston
6838	*Goodmoor Grange*	9/1937	11/1965	Cashmore, Newport
6839	*Hewell Grange*	9/1937	5/1965	Cohen, Kettering
6840	*Hazeley Grange*	9/1937	2/1964	Swindon Works
6841	*Marlas Grange*	9/1937	6/1965	Cohen, Morriston
6842	*Nunhold Grange*	9/1937	11/1964	Cashmore, Great Bridge
6843	*Poulton Grange*	10/1937	2/1964	Cohen, Morriston
6844	*Penyhdd Grange*	10/1937	4/1964	Cohen, Kettering
6845	*Paviland Grange*	10/1937	9/1964	Cashmore, Great Bridge
6846	*Ruckley Grange*	10/1937	9/1964	Birds, Bridgend
6847	*Tidmarsh Grange*	10/1937	12/1965	Buttigieg, Newport
6848	*Toddington Grange*	10/1937	12/1965	Cashmore, Newport
6849	*Walton Grange*	10/1937	12/1965	Cashmore, Newport
6850	*Cleeve Grange*	10/1937	12/1964	Cashmore, Newport
6851	*Hurst Grange*	11/1937	8/1965	Birds, Long Marston
6852	*Headbourne Grange*	11/1937	1/1964	Cashmore, Newport
6853	*Morehampton Grange*	11/1937	10/1965	Cohen, Kettering
6854	*Roundhill Grange*	11/1937	9/1965	Birds, Long Marston
6855	*Saighton Grange*	11/1937	10/1965	Cohen, Kettering
6856	*Stowe Grange*	11/1937	11/1965	Cashmore, Newport
6857	*Tudor Grange*	11/1937	10/1965	Cohen, Kingsbury
6858	*Woolston Grange*	12/1937	10/1965	Cohen, Kingsbury
6859	*Yiewsley Grange*	12/1937	11/1965	Buttigieg, Newport
6860	*Aberporth Grange*	2/1939	2/1965	Swindon Works
6861	*Crynant Grange*	2/1939	10/1965	Cohen, Kettering
6862	*Derwent Grange*	2/1939	6/1965	Birds, Bridgend
6863	*Dolywhel Grange*	2/1939	11/1964	Birds, Bridgend
6864	*Dymock Grange*	2/1939	10/1965	Cohen, Kingsbury
6865	*Hopton Grange*	3/1939	5/1962	Swindon Works
6866	*Morfa Grange*	3/1939	5/1965	Cashmore, Great Bridge
6867	*Peterston Grange*	3/1939	8/1964	Birds, Bridgend
6868	*Penrhos Grange*	3/1939	10/1965	Friswell, at Banbury shed
6869	*Resolven Grange*	3/1939	7/1965	Birds, Morriston
6870	*Bodicote Grange*	3/1939	9/1965	Cashmore, Great Bridge
6871	*Bourton Grange*	3/1939	10/1965	Cohen, Kettering
6872	*Crawley Grange*	3/1939	12/1965	Cashmore, Newport
6873	*Caradoc Grange*	4/1939	6/1964	Swindon Works
6874	*Haughton Grange*	4/1939	9/1965	Cashmore, Great Bridge
6875	*Hindford Grange*	4/1939	3/1964	Birds, Bridgend
6876	*Kingsland Grange*	4/1939	11/1965	Cashmore, Newport
6877	*Llanfair Grange*	4/1939	3/1965	Swindon Works
6878	*Longford Grange*	5/1939	11/1964	Buttigieg, Newport
6879	*Overton Grange*	5/1939	10/1965	Cohen, Kettering

Granges were extremely popular, and well suited to their work, even before re-draughting. 6869 *Resolven Grange* waits at Cardiff General, paired with one of the smaller 3,500-gallon Collett tenders. *HMRS/R. E. Lacey*

Below left. A Grange class 4-6-0 climbs the 1 in 66 gradient between Goodrington and Churston overlooking Torbay heading an Exeter–Kingswear train in the last days of BR steam a decade or so before the Dart Valley Railway took over the line. *T. E. Williams*

No 6865 *Hopton Grange* on Southern Region metals at Clapham Junction, and paired with a Collett 4,000-gallon tender, and since it was by then in BR ownership, a smokebox numberplate. *Lens of Sutton*

Tenders

As mentioned earlier, the further recycling of 43XX moguls resulted in second-hand tenders being paired with the new 68XX locomotives. The standard 3,500-gallon tender, with a coal capacity of 7 tons, ran on a 15ft 0in wheelbase, equally divided. When empty these tenders weighed only 18 tons 5 cwt. Water pick-up gear was a standard fitting. Tender framing was quite shallow on the Churchward design of tender, deepening only at the axle positions, although this was

No 6868 *Penrhos Grange* hauling an unusually light parcels train, tender-first, in the latter days of the GWR. *L. C. Jacks' Collection*

modified in later years, when some Granges were paired with the modified version. The 3,500-gallon water tank surrounded the coal space, extending to the front of the tender. Curiously, the tank on the right hand side was some 6in longer than on the left. Water scoop control handles were carried on the left, with the handbrake standard on the right.

During and after World War II Granges began to receive 4,000-gallon tenders of the type introduced by Collett in 1927. These larger tenders looked a much better balance with the 68XX 4-6-0s, with taller sides, but carried on the same 15ft 0in underframe, with deeper frame plates. The 4,000-gallon tenders were 4¼ tons heavier than the 3,500-gallon version when empty, but carried only 6 tons of coal.

Operations
As 'red' route engines, the Granges were restricted to similar routes worked by the Halls, but soon proved themselves to be excellent steaming, and they provided a smoother ride than other two-cylinder types. They developed an enviable reputation, and were very popular with GWR and BR enginemen, and improved still further following re-draughting during the early 1950s.

Allocations for the Granges covered almost any shed on the GWR and BR Western Region, from Old Oak Common to Penzance, and Fishguard to Birkenhead. Withdrawals began to take place in 1960, with some seeing out steam on the Western Region, and an average lifespan of between 24 and 29 years, although it is unfortunate that none was rescued for preservation. As with the Halls, the Granges were to have the names of country homes within the GWR's territory.

All 80 Grange class locomotives passed into BR hands in 1948, with a widespread distribution throughout the West Midlands and West Country. In 1950, three-quarters of the class could be found allocated to depots in the Newton Abbot, Bristol, and Wolverhampton Divisions. The majority were located at Newton Abbot – a picture which remained largely unchanged throughout the 1950s. But early in the 1960s 68XX class locomotives began disappearing at an ever-increasing rate, as diesel traction spread across the Western Region.

The Granges were among the most popular of the former Great Western two-cylinder 4-6-0s, and although they were reliable and successful in their original condition, their performance was improved still further in BR days. Like the Halls, Grange class locomotives were fitted with the Standard No 1 boiler, and jumper top blastpipe. Redraughting, with the removal of the jumper ring, sharpened the locomotives' performance, turning an already free-steaming boiler into an even better one.

Allocations of Grange class locomotives in November 1965, the Western Region's final year of steam operations, were as follows: 2A Tyseley (formerly 84E) 11, 2B Oxley (formerly 84B) 8, 81F Oxford 8, 85A Worcester 12, 88A Cardiff East Dock 5, 87F Llanelly 1.

No members of the class have been preserved.

The Manors

The Manor class was a lightweight 'blue' route 4-6-0, a smaller version of the Grange, as a replacement for 43XX 2-6-0s over routes where lighter axle loading was required. The first ten were announced in 1936, although they were not built until 1938, and a list of names was published in *The Railway Magazine*. The only change from the earlier published details to those of the locomotives as built was the dropping of the planned name *Ashley Manor* for No 7800 in favour of *Torquay Manor*

The design of the Manors continued Churchward's strict standardisation policies under C. B. Collett, where the company's intention was to provide 4-6-0 types to cover all duties where passenger tender locomotives were required. This demanded the final elimination of the typical British passenger 4-4-0. At the turn of the century these

suggestions may indeed have been revolutionary, but with hindsight it is seen as a logical development, allowing for the substantial influence of American locomotive design on Churchward's thinking.

Manor class locomotives were the first GWR 4-6-0s to be fitted with a smaller boiler than the Standard No 1. In the event, a completely new and much lighter boiler designated Standard No 14, was designed and fitted to the new locomotives. The considerable reduction in weight with the new design provided the Manors with a maximum axle load of only 17¼ tons on the leading coupled wheels. Using the reconditioned wheels and motion from withdrawn 2-6-0s, the Manor class locomotives had the same wheelbase as the larger Grange class, but were shorter in overall length.

Boiler details
The new Standard No 14 boilers built for Manor class were only about 75% of the physical size of the Standard No 1, and the overall length was some 2ft 4in shorter. A reduction in cylinder diameter from 18½in to 18in resulted in a nominal tractive effort of 27,340 lb. The end result was a boiler that did not match up to the traditions of performance expected from GWR design, while the evaporation rate that could be sustained in the original design was hampered by Churchward practices in the draughting

---

Manor class 4-6-0 – leading details

Wheel arrangement	4-6-0					Heating surface			
						– tubes	1285.50 sq ft		
Running numbers	7800–7829					– firebox	140.00 sq ft		
Lot numbers	316, 377					– total	1425.50 sq ft		
Built	1/1938–12/1950					– superheater	160.00 sq ft		
Withdrawn	4/1963–12/1965								
						Firebox			
Wheel diameter						– length o/s	8ft 8⅛in		
– coupled	5ft 8in					– width o/s	5ft 5½in		
– bogie	3ft 0in								
Wheelbase	27ft 1in					Grate area	22.1sq ft		
						Boiler pressure	225 lb/sq in		
Axle load	Tons Cwt	Tons Cwt	Tons Cwt			Tractive effort	27,340 lb		
	leading	driving	trailing						
– coupled	17 05	17 01	16 02			Fuel capacities			
	Tons Cwt					– coal	7 tons		
– bogie	18 10					– water	3,500 gallons		
Boiler						Weights			
– diameter	4ft 7⅞in tapering to 5ft 3in						*Full*		*Empty*	
– length	12ft 6in						*Tons* *Cwt*		*Tons* *Cwt*	
– tubes, small	158 × 2in o/d					– locomotive	68 18		63 06
– tubes, large	12 × 5⅛in o/d					– tender	40 00		18 05
– superheater elements	72 × 1in o/d								
Cylinders – bore × stroke	18in × 30in					– total	108 18		81 11

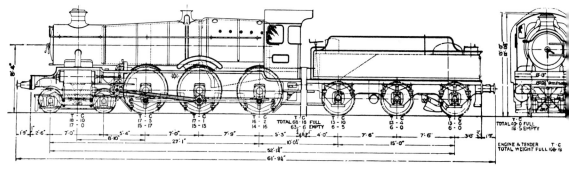

78XX Manor class 4-6-0

arrangements. But this boiler in its original form, with the consequent erratic effects on the locomotives' overall performance, remained unchanged for more than 13 years before any serious attempts were made to solve the problem. The locomotives were intended for and set to work on duties which would not have demanded maximum performance and it was not until more exhaustive testing was carried out into locomotive draughting in BR days, that the Manors' boiler inadequacies were cured.

Thirty-two Standard No 14 boilers were built between 1938 and 1950, with 158 2in tubes, and twelve 5⅛in diameter superheater flues, with 72 1in diameter elements. These latter were subsequently shortened, reducing the superheater heating surface from 190sq ft to 182.43sq ft. With the smaller No 14 boiler the firebox, although of standard design, was correspondingly reduced in size. The heating surface of the new firebox was just about 90% of that of the Standard No 1 boiler assembly, while that of the boiler proper had been drastically curtailed. At 22.1sq ft, the grate area was only 82% of that installed on locomotives with the Standard No 1 boiler, with the traditional layout of flat rear and sloping front half. At the front end the traditional Swindon arrangement was a burden to the performance of the Manors. The blastpipe and nozzle with which almost all earlier 4-6-0s were fitted had an internal diameter of 5¼in and was provided with the jumper top. In the Manor class with the much smaller boiler the jumper cap was retained, but a nominal reduction in blast nozzle diameter only was made, reducing it to 5⅛in, with a chimney choke diameter of 15in. Under later tests, this layout, when compared with that of the BR Standard Class 4 4-6-0, was shown to be behind the Manors' poor performance.

Frames, wheels and motion
The backbone of the locomotive and its running gear followed traditional Swindon design and construction practice, with the main plate frames terminating just in front of the leading axle, and behind the cylinders. The front extension bar frames were narrowed in from the normal 4ft 0in to 3ft 0in apart at the buffer beam. The usual generous size axleboxes were fitted to the coupled axles – 10in long, 8¾in diameter, and plain bearing type – with the locomotive weight supported by underhung leaf springs.

The 18in by 30in cylinders, with 10in diameter piston valves, were cast in two halves, including the smokebox saddle. As in the Grange class locomotives, the steamchest was larger than in previous practice, and offset slightly above the cylinders. (Centre-to-centre distance of the cylinders was 6ft 10in, against 5ft 10½in for the valves.) This new arrangement resulted in the raised footplating over the cylinders, at the front end of the locomotive. The 'straight' outside steampipes to the cylinders actually had a pronounced bend a few inches above the footplate where two sections of pipe joined before entering the smokebox. Valve gear was the standard Stephenson two-cylinder layout, with rocking shafts on either side transferring movement to and from the valve rod.

Brakes were steam-operated on the locomotive, with shoes carried behind the wheels, and sandpipes arranged to deliver ahead of the leading coupled wheels and behind the trailing pair. The Manors were the last of the two-cylinder designs to be fitted with the Churchward–DeGlehn bar frame bogie, supporting almost as much weight – 18½ tons at the front end – as the larger Grange class locomotives.

Detail features
Initially, the Manors were provided with

Grange type copper-capped chimneys, but following the re-draughting a new design of chimney, narrower than the predecessor, without smoke deflector was fitted. During World War II, the side windows of the cab were blanked-over on Nos 7800–7819, in common with many other types, but the windows were replaced soon after hostilities ceased. The usual GWR standard four-cone ejectors were conspicuous by their absence from the Manors; on other locomotives they were on the outside of the firebox, in front of the cab on the driver's side. From new, the Manors were equipped with AWS gear. The top headlamp bracket was fixed to the top of the smokebox door.

Another minor departure from normal practice was the provision of a much narrower footplate angle or valance, with the running boards supported at five points from the mainframes, and front frame extensions. The crosshead-driven pump was attached in the normal manner, hung from this valance, on the right-hand side. Collett's side window cab fitted to the Manors was still quite small in comparison with designs fitted to other companies' locomotives, and measured less than 5ft 0in long by 8ft 0in wide. The layout of controls for footplatemen followed standard Swindon practice.

From new, the Manors were paired with Churchward type 3,500-gallon tenders, carrying 7 tons of coal, and weighing some 40 tons when fully loaded. Like the Granges, these new lightweight 4-6-0s were paired with second-hand tenders from withdrawn 43XX moguls. Similarly, the Manors utilised wheels and motion from a pool of reconditioned parts taken from withdrawn 2-6-0s, most probably Nos 4302/7/12/16/19/21/22/25/42/43/49/55/60/69/79/82/89/99, 8301/63.

Construction

The locomotives were built to Diagram A9, under Swindon Lot Nos 316 and 377. The first 20 appeared in 1938 and 1939, with a further 10 constructed in 1950. The power class was D, and availability was over 'blue' routes. While the Granges were almost exactly the type of mixed traffic locomotive outlined by Churchward back in 1901, the Manors were equally the continuation of Churchward's design principles, with some minor variations. With their very light axle loads they were ideally suited to work on the Cambrian line in mid-Wales, where they first saw service in World War II.

Manor class 4-6-0 – building and withdrawal dates

Running No	Name	Built	Withdrawn	Disposal/scrapping
7800	Torquay Manor	1/1938	8/1964	Cashmore, Great Bridge
7801	Anthony Manor	1/1938	7/1965	Birds, Morriston
7802	Bradley Manor	1/1938	11/1965	Preserved
7803	Barcote Manor	1/1938	4/1965	Birds, Bridgend
7804	Baydon Manor	2/1938	9/1965	Cashmore, Newport
7805	Broome Manor	3/1938	12/1964	Cashmore, Great Bridge
7806	Cockington Manor	3/1938	11/1964	Cashmore, Great Bridge
7807	Compton Manor	3/1938	11/1964	Cashmore, Great Bridge
7808	Cookham Manor	3/1938	12/1965	Preserved
7809	Childrey Manor	4/1938	4/1963	Swindon Works
7810	Draycott Manor	12/1938	9/1964	Birds, Morriston
7811	Dunley Manor	12/1938	7/1965	Birds, Morriston
7812	Erlestoke Manor	1/1939	11/1965	Preserved
7813	Freshford Manor	1/1939	5/1965	Birds, Morriston
7814	Fringford Manor	1/1939	9/1965	Birds, Long Marston
7815	Fritwell Manor	1/1939	10/1964	Birds, Bridgend
7816	Frilsham Manor	1/1939	11/1965	Cashmore, Newport
7817	Garsington Manor	1/1939	6/1964	Birds, Risca
7818	Granville Manor	1/1939	1/1965	Cashmore, Great Bridge
7819	Hinton Manor	2/1939	11/1965	Preserved
7820	Dinmore Manor	11/1950	11/1965	Preserved
7821	Ditcheat Manor	11/1950	11/1965	Preserved
7822	Foxcote Manor	12/1950	11/1965	Preserved
7823	Hook Norton Manor	12/1950	7/1964	Cashmore, Great Bridge
7824	Iford Manor	12/1950	11/1964	Cashmore, Great Bridge
7825	Lechlade Manor	12/1950	5/1964	Birds, Risca
7826	Longworth Manor	12/1950	4/1965	Birds, Bynea
7827	Lydham Manor	12/1950	10/1965	Preserved
7828	Odney Manor	12/1950	12/1965	Preserved
7829	Ramsbury Manor	12/1950	12/1965	Cashmore, Newport

Operations

Initial allocations of the first batch, Nos 7800–7809 built in 1938, were made to cover through Newcastle trains on the Banbury–Cheltenham–Swansea section, on secondary main line duties. During World War II a number of Manors were transferred to Oswestry, and began working over the 'blue' routes of the Cambrian lines in mid-Wales. Before the arrival of the new locomotives, traffic on the Cambrian section was handled by the elderly Dean designed 4-4-0s, supplemented by the hybrid 3200 class 4-4-0s, nicknamed 'Dukedogs', constructed in 1936. When the Manors were finally set to work on the 'blue' routes, there was some small reaction to the running of 4-6-0s on these lines, where previously only 4-4-0s and 2-6-0s had handled the regular work.

By the end of World War II, in the last couple of years of the GWR's independent existence, Manor class locomotives were much more widely available. Bristol and Banbury each had five, with three at Oswestry, two at Gloucester, two at Machynlleth, and one each at Wolverhampton, Leamington and Croes Newydd, in 1947. In the early 1950s, no fewer than 11 were stabled in Devon and Cornwall, displacing yet more 4-4-0s on banking and pilot duties, with the Cambrian contingents equally strong.

Nos 7820–7829, built in 1950, were completely new and did not use any reconditioned components from withdrawn moguls, as the earlier locomotives had done. The locomotives built in BR days came out with smokebox numberplates, and a new power

Top right. A broadside view of No 6802 *Bampton Grange* in GWR unlined green livery, with the 'shirtbutton' emblem on the tender sides, and sporting the cast-iron chimney originally fitted to the first Grange class Nos 6800–6803. *Locomotive & General Railway Photographs*

Centre right. No 7804 *Baydon Manor* in GWR unlined green livery with the 'shirtbutton' emblem. Comparing this view with that of the Grange class reveals the similarity in outward appearance between the two. *Locomotive & General Railway Photographs*

Right. The Manor class, was not a happy design when it first appeared. Due to the hostilities of 1939–45 the problems were not resolved until after nationalisation. Despite the side window cab, and other detail improvements, the locomotives were undoubtedly of Churchward descent. No 7816 *Frilsham Manor*, like its sisters, was a locomotive transformed after re-draughting, and a very popular type in later years. *G. W. Sharpe*

Grange class 4-6-0 No 6873 *Caradoc Grange* pilots King class 4-6-0 No 6017 on a Penzance–Birmingham train at Aller Junction approaching Newton Abbot in September 1959. *Derek Cross*

classification, 5MT. They were built with the traditional Churchward front end, and consequently retained the poor performance characteristics of the original locomotives. In 1951 S. O. Ell's thorough testing of the new BR Standard Class 4MT 4-6-0, with a comparative examination of No 7818 *Granville Manor* revealed the nature of the problem. The proportions of the chimney and blastpipe of the original design under test could manage a maximum evaporation rate of only 10,000 lb/hour. The simple expedient of reducing the blastpipe jumper ring and reducing the nozzle diameter to 4⅝in resulted in the maximum steam rate being more than doubled to 20,400 lb/hour. It is doubtful whether the Manors without the benefit of S. O. Ell's testing at Swindon would have improved their reputation as unreliable and inferior steamers. As it was the changes that were made and subsequently included on other GWR types transformed the Manors into possibly the most successful design of their type.

At the time of building the second batch of Manors in 1950, most of these small 4-6-0s could be found at Newton Abbot, Laira, Banbury, and Leamington, with a handful on the Central Wales line, working out of Oswestry. By 1960 the allocations of individual locomotives had changed but the general disposition of the class as a whole throughout the region was the same, although rather more of their number could be found in South Wales. Withdrawal was a fairly rapid process, with only one locomotive going in 1963, 10 in 1964, and the remaining 19 in 1965. Thirteen were withdrawn between September and December of that year, marking the end of steam traction on the Western Region.

Sadly perhaps, the second production run of Manors had their careers cut short by the onset of diesel traction. The first of the class to succumb to the breaker's torch was No 7809 *Childrey Manor*, which was withdrawn in April 1963, while all were gone by the time the Western Region was fully dieselised at the end of 1965. The popularity of the Manors, revitalised with their improvements of the 1950s, may be evidenced by the fact that no fewer than nine members of this 30-strong class have been rescued for preservation.

CHAPTER SIX

BEGINNINGS OF CHANGE – THE HAWKSWORTH ERA

C. B. Collett had been CME of the Great Western Railway for some 20 years, and had to all intents and purposes continued the ideology and traditions of Churchward. During Collett's time only two new designs of two-cylinder tender locomotive were produced, while another was simply a Churchward design of 1901, whose introduction had merely been delayed by a few years.

Twenty years after Churchward there had been little by way of the degree of change to what the great man had brought about, although such events were soon to take place. Presiding over the GWR's motive power policy during World War II and in its final years of independence was F. W. Hawksworth, who quickly began to introduce a number of non-traditional features into the design and construction of Swindon's locomotives. He also approached the idea of introducing non-steam traction into the Great Western main line stock, and here as in so many other cases the GWR took a

very different line from that of other British railway companies.

Hawksworth took on the role of CME of the Great Western in 1941, having previously been the company's chief locomotive draughtsman, with responsibility for testing. In his approach to matters of design and construction he was on a similar wavelength to Churchward, as compared with his immediate predecessor, and his first two-cylinder design showed ample evidence of this. The 6959 or Modified Hall class was a group of 71 locomotives which while retaining some features of the original Hall class were very different in many ways. The 6959 class was built in four Lots – Nos 350, 366, 368, and 376, between 1944 and 1950, and the first of Lot No 368, No 6991 was the

Hawksworth introduced a variety of changes into GWR locomotive design policy, with his first essay appearing in the shape of the Modified Hall class. No 7910 *Hown Hall* is paired with the Hawksworth self-trimming tender. *G. W. Sharpe*

very first GWR type to be built new with a smokebox numberplate.

While it would be true to say that Hawksworth carried out a number of experiments on the Modified Halls, and undoubtedly improved the breed, the later County class was something completely new to GWR thinking. The Counties were, it has been said, the final development of the much acclaimed Saint class 4-6-0, but although this may be true in principle there were many differences in design and operation. The Counties were the GWR's final two-cylinder mixed traffic design, not departing so much from Churchward and Swindon practice and traditions, as improving them. It is true to say that the initial design revealed some shortcomings, but these were soon resolved, and the class represented the most powerful locomotive of its type in service with the Great Western. In terms of power the Counties were rated somewhere between Halls and Castles; classified as 6MT by British Railways, this put the Counties in the same league as Stanier's Royal Scot 4-6-0s on the LMS. The earlier Modified Halls were still 5MT in the BR scheme.

Modified Hall Class 4-6-0

This new class of two-cylinder 4-6-0 began life as part of some rethinking of GWR boiler design and proportions, resulting from the experience of unreliable performance during World War II with some of the Churchward designs. GWR locomotives were designed for performance on a fairly narrow range of fuel types, and some changes were necessary to improve their

Modified Hall class 4-6-0 – leading details

Class	6959, Modified Hall
Wheel arrangement	4-6-0
Running numbers	6959–7929
Lot numbers	350, 366, 368, 376
Built	3/1944–11/1950
Withdrawn	1/1963–12/1965

Wheel diameter	
– coupled	6ft 0in
– bogie	3ft 0in
Wheelbase	27ft 2in

Axle load	Tons Cwt leading	Tons Cwt driving	Tons Cwt trailing
– coupled	19 00	19 05	19 05
	Tons Cwt		
– bogie	18 06		

Boiler	
– diameter	4ft 10$\frac{13}{16}$in tapering to 5ft 6in
– length	14ft 10in
– tubes, small	145 × 2in o/d
– tubes, large	21 × 5$\frac{1}{8}$in o/d
– superheater elements	84 × 1$\frac{1}{4}$in o/d

Cylinders – bore × stroke	18$\frac{1}{2}$in × 30in

Heating surface	
– tubes	1582.6 sq ft
– firebox	154.9 sq ft
– total	1737.5 sq ft
– superheater	314.6 sq ft

Firebox	
– length o/s	9ft 0in
– width o/s	5ft 9in/4ft 0in

Grate area	27.07 sq ft

Boiler pressure	225 lb/sq in

Tractive effort	27,275 lb

Fuel capacities	
– coal	6 tons (Hawksworth tenders – 7 tons)
– water	4,000 gallons

Weights	Full Tons Cwt	Empty Tons Cwt
– locomotive	75 00	69 00
– tender	46 14	22 10
– total	121 14	91 10

6959 Modified Hall class 4-6-0

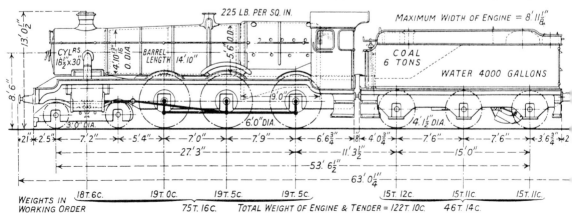

operations on lower grade fuels. This potential fallibility in Churchward's designs was emphasised during the 1948 Interchange Trials, when the Modified Hall class selected for comparative testing turned in some poor performances against other types, even on its own territory.

Boiler design on the GWR had remained largely unchanged from Churchward's day, save for the wider application of superheating. It was in superheating that the most significant changes were made to boiler design, fitting in a three-row 21-flue arrangement, instead of the previous standard two-row 14-flue layout.

Mechanically the Modified Halls were substantially and significantly altered from their forerunners, with simpler design and construction features, especially in framing and bogies. An important trend in locomotive design at this time was moving towards simplicity, with the emphasis on lower construction and maintenance costs. On the GWR, the new 4-6-0s were a first step in that direction, with a number of dimensional differences between Hall and Modified Hall, with the principal features listed opposite.

Boiler details

The boiler itself was essentially a Standard No 1, with the same smokebox and firebox dimensions and features as the earlier Hall class. The same draughting arrangements applied too, with the jumper top blastpipe, but the number and size of tubes and superheater flues was much different. In the Modified Halls an additional row of seven 5⅛in diameter superheater tubes was installed with a larger header in the smokebox. The 84 elements were also larger, with an outside diameter of 1¼in compared with only 1in for the earlier standard two-row superheaters. In 1947, the superheater heating surface was reduced to 303.6sq ft, by shortening the elements, and later still in British Railways days the heating surface was further reduced, to 295sq ft. This higher degree of superheat, with steam temperatures up to around 600°F, was achieved with the new standard three-row superheater, compared with the low to medium superheat possible with the older standard two-row variety.

The front end of the locomotive's steam raising plant was essentially unchanged, save for minor alterations to accommodate the larger superheater. Similarly, the firebox design remained largely unaltered, except for two experiments carried out on Nos 6965 and 6967. The former was fitted with a welded steel firebox, whilst the latter had a hopper type ashpan and rocking grate.

Frames, wheels and motion

Perhaps the most notable and noticeable improvements and changes made by Hawksworth in the Modifed Hall was in the design and construction of the locomotive's mainframes. Here 1¼in thick mild steel plates were used, stayed at a distance of 4ft 1in apart, but in the new locomotives they ran from the rear dragbeam, to the front bufferbeam – the full length of the locomotive. This demanded a major change in cylinder layout, design and installation. Previous Swindon standard practice had been to cast the cylinders in two halves, each complete with part of the smokebox saddle, but in the new Modified Hall class it was necessary to bolt the cylinders to the machined seating on the outside face of the mainframes. Between the frames, a steel cross-stay was extended upwards to form the smokebox saddle. This new design of mainframe assembly eliminated the traditional Swindon practice of part plate, and part bar frame, and was later used in the County class 4-6-0s.

Although the overall weight of the Modified Halls showed an increase over the Hall class, the load on the bogie was reduced by 4 cwt. The loads on the coupled axles were increased by 8 cwt, 6 cwt, and 6 cwt respectively. The leading bogie was completely redesigned with an all plate frame construction commonly seen on other railways, but retaining the standard GWR 10-spoke, 3ft 0in diameter wheels. Another feature of the altered mainframe construction was the very different appearance of the front end of the locomotive. In the Modified Halls the front portion of the frames protruded above the front footplate, with the supporting cross stay/smokebox saddle easily visible.

The design of the 18½in × 30in cylinders was typically Swindon with the exception of the layout of the castings, as was the standard inside Stephenson valve gear. A characteristic operating feature of GWR two-cylinder locomotives was a fairly pronounced longitudinal motion, in some

The new design of plate frames and bogie can clearly be seen on No 7920 *Coney Hall*, along with mechanical lubrication. Many standard features were retained, including the ejectors mounted on the firebox side, blower pipe fitting, and arrangement of front end lubricator pipes. *G. W. Sharpe*

measure perhaps due to the settings used in the motion, and was reportedly even more pronounced in the new locomotives. In the first major modification to Churchward's traditional front end layout, steam and exhaust pipe routings were altered. On one test run with a Modified Hall during the 1948 Interchange Trials, the internal steam-pipes leaked badly, and whatever the merits or demerits of the Churchward arrangement, such incidents did not occur in his design.

Above the frames the new locomotives were still almost pure Churchward, except for the Collett side window cab. Most plate-work was still rivetted, although welding techniques were being introduced and applied in a variety of ways. While the Modified Halls were fitted with standard drum-head pattern smokeboxes, rivetted-up, later changes and repairs procedures resulted in welded joints at the smokebox door ring and boiler joints on some types. There is some evidence to suggest that such joints were welded on a number of 43XX 2-6-0s.

At the first boiler lagging band, and just behind the leading wheel splasher, the lagging was cut away to accommodate the bolting faces of the boiler saddles. On the fireman's side, the fire iron compartment alongside the firebox on the footplate was fitted from new, as also was the GWR's standard ATC gear.

The three coupled wheel splashers, despite the simpler construction practices, still had brass raised beadings in deference to long-standing tradition, like the copper-capped chimneys. In all the designs under discussion the centre splasher was wider, to leave room for the reversing gear reach rod passing behind the nameplate. Although GWR footplates were on the whole fairly clean and uncluttered in layout, without such items as sandbox fillers, mechanical lubricators and the like, a long sandbox operating rod reached forward from the cab to just in front of the valve rocker covers. Both leading sandboxes were operated in this way, with the box on the left side worked via a cross-shaft. The sandboxes were placed in the normal position, with delivery pipes ahead of the leading wheels, and behind the trailing pair.

Lubrication in the traditional GWR form was hydrostatic, or sight feed, where the

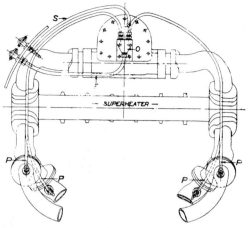

Arrangement of Lubricator Pipes in Smokebox.

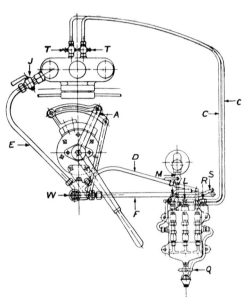

Arrangement for Controlling Cylinder Lubrication
by Regulator.

A	Actuating arm.	Q	Drain cock.
C	Steam pipe.	R	Filling plug.
D	Oil pipe.	S	Oil-feed pipe.
E	Steam pipe.	T	Steam-pipe cocks.
F	Auxiliary steam pipe.	U	Cock.
J	Cock.	W	Regulator-controlled
L	Spare feed handle.		valve.
M	Warming cock.		

driver controlled the flow of oil to valves and pistons by counting the number of drops passing through the sight glass. It was required that 15 spots of oil were to pass through the sight glass every two minutes. By this means, the GWR lubricated all its steam locomotives until various experiments with mechanical lubricators were tried again in the 1930s. These were later removed from the Hall class locomotives to which they were fitted, and it was not until the Modified Halls appeared that the method was used again, but more extensively. Even so, not all the Modified Halls were provided with this equipment. Those locomotives which carried mechanical lubricators had them fitted on the right-hand side on the running plate, in front of the leading coupled wheel splasher. In addition locomotives fitted with mechanical lubrication carried a gauge in the cab which read 'oil' or 'no oil' depending on whether the regulator was closed or open. Some enginemen believed that with the regulator just open, as when coasting, the indication of the gauge might show 'oil' when only steam was being delivered to the cylinders and valves.

While there were two different schools of thought on the most effective means of lubrication, the Modified Halls and certain other classes fitted with mechanical lubricators earned a reputation as very speedy machines.

Boiler changes produced a number of detail variations, including some carrying the old Swindon Standard No 3 superheater, with two rows of flues instead of three. A different diagram was issued for locomotives fitted with this type of superheater. Chimneys were largely the Grange type, with smoke deflector, although a number carried a plain type of chimney without a copper top; a third variation was a taller than standard chimney.

Construction

Seventy-one 6959 class locomotives were turned out by Swindon in four lots between March 1944 and November 1950. Most built during the British Railways period, with only 22 as purely GWR builds. Construction overall was a little erratic, with the first dozen appearing in 1944 under Lot No 350, with a gap of just over two years before the next batch began to appear, from October

No 7925 *Westol Hall* never carried GWR livery, being built in 1950 as a BR Western Region locomotive. But despite detail differences it was still a typical GWR type. *Lens of Sutton*

1947 onwards. None of the 6959 class was fitted for oil burning.

The Modified Halls, like their predecessors, were power class D, route availability 'red'. The three diagrams issued were: A14–6959 class, with three-row superheater; A15–6959 class, with two-row superheater; A25–6959 class, with Hawksworth pattern tender.

On the fireman's side the Modified Halls had the fire iron tunnel alongside the firebox as standard practice, while for No 7923 *Spoke Hall* a Collett 4,000-gallon tender was used. *Lens of Sutton*

Tenders

Another new feature of the early Hawksworth period was the introduction of a different tender design, which first appeared attached to No 6974 *Bryngwyn Hall*. The design had already been seen with the new County class 4-6-0s from the autumn of 1945. It was very different to previous GWR practice, with a redesigned underframe and straight sides to the tender tank and coal space, but was of the same basic dimensions as Collett's 4,000-gallon tender. The wheelbase was 15ft 0in, equally divided, with 4ft 0in wheels, and main framing similar in arrangement to that adopted on the LMS and other railways. This was not the last unusual tender to be attributed to Swindon's

Modified Hall class – building and withdrawal dates

Running No	Name	Built	Withdrawn	Scrapping/disposal
6959	Peatling Hall	3/1944	12/1965	Cashmore, Newport
6960	Raveningham Hall	3/1944	6/1964	Preserved
6961	Stedham Hall	3/1944	9/1965	Cashmore, Newport
6962	Soughton Hall	4/1944	1/1963	Swindon Works
6963	Throwley Hall	4/1944	7/1965	Birds, Long Marston
6964	Thornbridge Hall	5/1944	9/1965	Ward, Sheffield
6965	Thirlestaine Hall	7/1944	10/1965	Cashmore, Newport
6966	Witchingham Hall	5/1944	9/1964	Birds, Risca
6967	Willesley Hall	8/1944	12/1965	Cashmore, Newport
6968	Woodcock Hall	9/1944	9/1963	Cashmore, Newport
6969	Wraysbury Hall	9/1944	2/1965	Swindon Works
6970	Whaddon Hall	9/1944	6/1964	Birds, Risca
6971	Athelhampton Hall	10/1947	10/1964	Cashmore, Great Bridge
6972	Beningborough Hall	10/1947	3/1964	Birds, Bridgend
6973	Bricklehampton Hall	10/1947	8/1965	Cashmore, Newport
6974	Bryngwyn Hall	10/1947	5/1965	Birds, Risca
6975	Capesthorne Hall	10/1947	12/1963	Slag Reduction Co, Briton Ferry
6976	Graythwaite Hall	10/1947	10/1965	Cashmore, Great Bridge
6977	Grundisburgh Hall	11/1947	12/1963	Birds, Risca
6978	Haroldstone Hall	11/1947	7/1965	Birds, Swansea
6979	Helperly Hall	11/1947	2/1965	Birds, Long Marston
6980	Llanrumney Hall	11/1947	10/1965	Cashmore, Great Bridge
6981	Marbury Hall	2/1948	3/1964	Birds, Bridgend
6982	Melmerby Hall	1/1948	8/1964	Cashmore, Newport
6983	Otterington Hall	2/1948	8/1965	Cashmore, Newport
6984	Owsden Hall	2/1948	12/1965	Preserved
6985	Parwick Hall	2/1948	9/1964	Cashmore, Newport
6986	Rydal Hall	3/1948	4/1965	Birds, Bridgend
6987	Shervington Hall	3/1948	9/1964	Birds, Bridgend
6988	Swithland Hall	3/1948	9/1964	Birds, Bridgend
6989	Wightwick Hall	3/1948	6/1964	Preserved
6990	Witherslack Hall	4/1948	12/1965	Preserved
6991	Acton Burnell Hall	11/1948	12/1965	Cashmore, Newport
6992	Arborfield Hall	11·1948	6/1964	Birds, Risca
6993	Arthog Hall	12/1948	12/1965	Cashmore, Newport
6994	Baggrave Hall	12/1948	11/1964	Cashmore, Great Bridge
6995	Benthall Hall	12/1948	3/1965	Birds, Risca
6996	Blackwell Hall	1/1949	10/1964	Steel Supply Co, Briton Ferry
6997	Bryn-Ivor Hall	1/1949	11/1964	Birds, Bridgend
6998	Burton Agnes Hall	1/1949	12/1965	Preserved
6999	Capel Dewi Hall	2/1949	12/1965	Cashmore, Newport
7900	Saint Peter's Hall	4/1949	12/1964	Friswell, Banbury
7901	Doddington Hall	3/1949	2/1964	Hayes, Bridgend
7902	Eaton Mascot Hall	3/1949	6/1964	Swindon Works
7903	Foremarke Hall	4/1949	6/1964	Preserved
7904	Fountains Hall	4/1949	12/1965	Cashmore, Newport
7905	Fowey Hall	4/1949	5/1964	Cashmore, Great Bridge
7906	Fron Hall	12/1949	3/1965	Swindon Works
7907	Hart Hall	1/1950	12/1965	Cashmore, Newport
7908	Henshall Hall	1/1950	10/1965	Cashmore, Great Bridge
7909	Heveningham Hall	1/1950	11/1965	Cashmore, Newport
7910	Hown Hall	1/1950	2/1965	Swindon Works
7911	Lady Margaret Hall	2/1950	12/1965	Birds, Risca
7912	Little Linford Hall	3/1950	10/1965	Cashmore, Great Bridge
7913	Little Wyrley Hall	3/1950	3/1965	Birds, Bynea
7914	Lleweni Hall	3/1950	12/1965	Cashmore, Newport
7915	Mere Hall	3/1950	10/1965	Cashmore, Great Bridge
7916	Mobberley Hall	4/1950	12/1964	Swindon Works
7917	North Aston Hall	4/1950	8/1965	Birds, Bynea
7918	Rhose Wood Hall	4/1950	2/1965	Cohen, Swansea
7919	Runter Hall	5/1950	12/1965	Cashmore, Newport
7920	Coney Hall	9/1950	6/1965	Cohen, Swansea
7921	Edstone Hall	9/1950	12/1963	Birds, Risca
7922	Salford Hall	9/1950	12/1965	Cashmore, Newport
7923	Speke Hall	9/1950	6/1965	Cohen, Swansea
7924	Thornycroft Hall	9/1950	12/1965	Cashmore, Newport
7925	Westol Hall	10/1950	12/1965	Cashmore, Newport
7926	Willey Hall	10/1950	12/1964	Buttigieg, Newport
7927	Willington Hall	10/1950	12/1965	Preserved
7928	Wolf Hall	10/1950	3/1965	Swindon Works
7929	Wyke Hall	11/1950	8/1965	Birds, Bridgend

Nos 6959 to 6970 were put into traffic un-named, a continuation of the austerity imposed on Nos 6916 to 6958, already noted. They received nameplates between January 1946 and September 1948, and until the nameplates were fitted the words 'Hall Class' were painted on the middle driving wheel splashers.

design team, since a few years later extensive use was made of aluminium in a one-off prototype. With a full load of 7 tons of coal, and 4,000 gallons of water, the new Hawksworth type weighed 49 tons.

The County class 4-6-0 – Hawksworth's 'guinea pigs'

It has been suggested that these locomotives were the final mixed traffic development of Churchward's prototype 4-6-0s No 98 and No 100, and the Saint class. There were some similarities, but a great many more differences – the end result was a locomotive whose appearance received a great deal of criticism, much of it unjustified. True, there had been much to admire in Churchward's locomotives, but even his designs received a considerable degree of misplaced criticism on their first arrival.

Hawksworth's new locomotive design was a long time in preparation, since he had taken office during World War II, and was engaged for some time on work for the war effort. Among those projects was the construction of large numbers of the Stanier Class 8F 2-8-0s for the Ministry of Supply. Changes in construction practice also began to have greater effect, with a more widespread use of welding as against rivetting, and a desire to improve the accessibility of working parts for repairs and maintenance. The role of the chief mechanical engineer had changed, with Hawksworth's position carrying much less authority than some of his predecessors, and he reported to the General Manager, Sir James Milne, instead of directly to the board. A result of this changed position, particularly when considering post-war locomotive developments, was the

greater interest taken by members of the board in locomotive matters.

The Locomotive Committee of the GWR, in formulating post-war policies towards the end of the hostilities, was examining the feasibility of constructing a pacific type for main line express work. However, the restrictions placed upon the company by wartime conditions, enabled the Ministry of Supply to stop work on the development of any new express passenger type. This was a disappointment, and one which was to lead ultimately, if a little ironically, to the production of the 1000 class Counties. Although the official position placed an embargo on the construction of a new GWR Pacific, some of the features were incorporated in the new 4-6-0, which began to take shape in 1944. At the same time, a number of alternative designs were discussed for the new 4-6-0, one of which included the fitting of outside Walschaerts valve gear. This particular alternative also included higher running boards or footplating, similar in appearance to the later British Railways Standard series locomotives, to give clearance and greater accessibility to the wheels and motion.

While the outside valve gear did not appear on No 1000 when it finally emerged from Swindon Works, the 6ft 3in coupled wheels and 280 lb/sq in boiler pressure were two features that would have been used in the projected pacific type. Originally it had been suggested that the Castle class boiler with some modifications would have been suitable, but its weight would have restricted the new locomotives' route availability. In the end, some advantage was taken of the experience of building large numbers of LMS Class 8F 2-8-0s for the war effort, and in particular, the jigs used in the construction of their boilers.

1000 County class 4-6-0

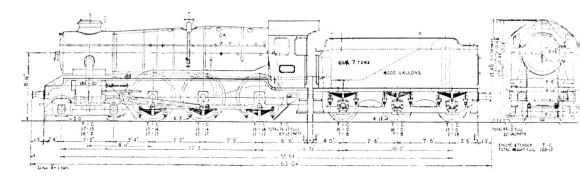

County Class 4-6-0 – leading details

Class	1000 Class, County
Wheel arrangement	4-6-0
Designer	F. W. Hawksworth
Running numbers	1000–1029
Lot numbers	354, 358
Built	8/1945–4/1947
Withdrawn	9/1962–11/1964
Length o/a	63ft $0\frac{1}{4}$in
Width o/a	8ft $11\frac{1}{8}$in
Height o/a	13ft 4 $\frac{31}{32}$in
Wheel diameter	
– coupled	6ft 3in
– bogie	3ft 0in
Wheelbase	27ft 3in

Axle load	*Tons*	*Cwt*	*Tons*	*Cwt*	*Tons*	*Cwt*
	leading		driving		trailing	
– coupled	19	14	19	14	19	14
	Tons	*Cwt*				
– bogie	17	15				

Boiler	
– type	Standard No 15
– diameter	5ft 0in tapering to 5ft $8\frac{3}{8}$in
– length	12ft $7\frac{3}{16}$in
– tubes, small	198 × $1\frac{3}{4}$in o/d
– tubes, large	21 × $5\frac{1}{8}$in o/d
– superheater elements	84 × $1\frac{1}{4}$in o/d
Cylinders – bore × stroke	$18\frac{1}{3}$in × 30in
Heating surface	
– tubes	1545.00 sq ft
– firebox	169.00 sq ft
– total	1714.00 sq ft
– superheater	254.00 sq ft
Firebox	
– length o/s	9ft 9in
– width o/s	5ft $10\frac{7}{8}$in
Grate area	28.84 sq ft
Boiler pressure	280 lb/sq in (later reduced to 250 lb/sq in)
Tractive effort	32,580 lb (at 280 lb/sq in)
Fuel capacities	
– coal	7 tons
– water	4,000 gallons

Weights	*Full*		*Empty*	
	Tons	*Cwt*	*Tons*	*Cwt*
– locomotive	76	17	69	13
– tender	49	00	22	14
– total	125	17	92	07

Boilers

The new Standard No 15 boiler was more than 2ft 0in shorter than the GWR Standard No 1, and housed 1¾in diameter tubes, compared with the usual 2in diameter, and a three-row superheater was installed in the conventional 5⅛ diameter flues. The superheater itself was later modified, shortening the 84 elements by 6in, and reducing the heating surface to 248sq ft. A higher degree of superheat was introduced to the GWR as an important departure from traditional Swindon practice on the Counties and Modified Halls, but with the former Hawksworth went even further and raised the boiler pressure to 280 lb/sq in. Such a high working pressure was almost unprecedented in normal practice on a British railway, although similar changes had been introduced by O. V. S. Bulleid on the Southern Railway. In the Counties the experiment did not last, and boiler pressure was later reduced to 250 lb/sq in.

The front end was substantially altered, with the inclusion of a double blastpipe and chimney on the pioneer locomotive of the class, in a longer than normal smokebox. At 7ft 0in, it was almost 12in longer than the previous Swindon standard, although construction followed the normal pattern. The pronounced dishing in the smokebox door was less obvious in the new design, and the diameter was greater at 5ft 0in, to mate with the boiler ring, and a correspondingly larger door. The double chimney of No 1000 was not repeated on subsequent locomotives, which had single chimneys, although redraughting experiments in BR days produced a return to the double blastpipe arrangement.

A larger firebox built in the traditional manner from copper provided 169sq ft of heating surface, greater even than that of the Castle class, and a new hopper type ashpan was fitted. The grate itself, with a sloping forward section and flat rear, was again conventional Swindon practice, and in common with changes introduced on the Modified Halls, a sloping front plate was used in this firebox design. One problem in the Counties was the firebox smokeplate, whose function was to direct secondary air over the firebed. In the new locomotives this tended to dip down, making firing to the front of the box a little difficult – footplate crews introduced various unofficial 'remedies' to correct this fault. When Swindon learned of these 'remedies' they were stopped, but the firing problem remained and was never officially cured.

The double blastpipe and chimney was only fitted for experimental purposes, and subject to early scrutiny by the locomotive testing staff at Swindon. Some character-

County class 4-6-0 No 1000 *County of Middlesex* was originally the only example to carry a double chimney. Seen here in immaculate condition, with speedometer drive from the trailing coupled axle. *Lens of Sutton*

istics of the locomotive provided indifferent performance, and the pronounced fore-and-aft motion typical of GWR two-cylinder types was exaggerated in the Counties.

Frames, Wheels and Motion
The County class locomotives were fitted with the same design of main frame assembly as the Modified Halls – entirely constructed of mild steel plate, stayed at 4ft 1in apart throughout. Main axleboxes were as seen on the 6959 class, generous plain bearing types, while the coupled wheels at 6ft 3in diameter with 21 spokes were non-standard, but were the same as those that would have been fitted to the projected Pacific. Underhung, uncompensated leaf springs provided the main suspensions while the plate frame bogie – the same as that fitted to the Modified Halls – had independent suspension for its 3ft 0in diameter wheels.

The 18½in × 30in cylinders, with 10in diameter piston valves were cast separately and bolted to the outside face of the frames. A cast-steel stretcher under the smokebox provided the dual function of main frame stay and smokebox support saddle. The piston valves had a full travel of 7½in. Compromise was reached in the design of the

valve motion itself, and despite the possible introduction of outside Walschaerts valve gear the standard Stephenson gear was fitted. Outside Walschaerts gear did appear on a GWR design eventually, but on Hawksworth's 1500 class 0-6-0 tank engines.

Details
While the major differences in the locomotive itself were in the steam-raising plant, a number of changes were made to other standard arrangements. At the smokebox joint, the lubricator pipes emerged on the right-hand side only, under their traditional cover. Another more obvious change was the single splasher covering all coupled wheels, with straight nameplates. On the fireman's side, the tool compartment on the left-hand side of the firebox was now a standard feature. Some difficulty was experienced with clearance for the reach rod behind the nameplates on the right-hand side. On this side it was not possible to fit the nameplate to the splasher, and it was made free-standing on a sheet steel support immediately in front of the reversing rod.

The cab was the same size as that fitted to the Castle class, and obviously much roomier by comparison with the earliest two-cylinder locomotives. The cab roof was the highest part of the locomotive, at 13ft 4½in above rail level, while the 16½in high single chimney was 2in lower, with the boiler centre line pitched at 8ft 11in. The footplate

In August 1958 No 1012 *County of Denbigh* was at Oxford shed, ready to work back to Swindon. Speedometer gear is fitted. The double chimney was provided on this locomotive in September 1957. *L. C. Jacks*

County class 4–6–0 – Building and withdrawal dates

Running No	Name	Built	Withdrawn	Scrapped at
1000	County of Middlesex	8/1945	7/1964	Cashmore, Newport
1001	County of Bucks	9/1945	5/1963	Cashmore, Newport
1002	County of Berks	9/1945	9/1963	Ward, Sheffield
1003	County of Wilts	10/1945	10/1962	Cashmore, Newport
1004	County of Somerset	10/1945	9/1962	Cashmore, Newport
1005	County of Devon	11/1945	6/1963	Cashmore, Newport
1006	County of Cornwall	11/1945	9/1963	Cooper, Sharpness
1007	County of Brecknock	12/1945	10/1962	King, Norwich
1008	County of Cardigan	12/1945	10/1963	Cashmore, Newport
1009	County of Carmarthen	12/1945	2/1963	Swindon Works
1010	County of Carnarvon*	1/1946	7/1964	Cashmore, Newport
1011	County of Chester	1/1946	11/1964	Cashmore, Newport
1012	County of Denbigh	2/1946	4/1964	Cashmore, Newport
1013	County of Dorset	2/1946	7/1964	Cashmore, Newport
1014	County of Glamorgan	2/1946	4/1964	Cashmore, Newport
1015	County of Gloucester	3/1946	12/1962	Cashmore, Newport
1016	County of Hants	3/1946	9/1963	Ward, Sheffield
1017	County of Hereford	3/1946	12/1962	Ward, Sheffield
1018	County of Leicester	3/1946	9/1962	King, Norwich
1019	County of Merioneth	4/1946	2/1963	Cashmore, Great Bridge
1020	County of Monmouth	12/1946	2/1964	Hayes, Bridgend
1021	County of Montgomery	12/1946	11/1963	Hayes, Bridgend
1022	County of Northampton	12/1946	10/1962	Ward, Sheffield
1023	County of Oxford	1/1947	3/1963	Swindon Works
1024	County of Pembroke	1/1947	4/1964	Swindon Works
1025	County of Radnor	1/1947	2/1963	Cashmore, Great Bridge
1026	County of Salop	1/1947	9/1962	Ward, Sheffield
1027	County of Stafford	3/1947	10/1963	Cooper, Sharpness
1028	County of Warwick	3/1947	12/1963	Birds, Risca
1029	County of Worcester	4/1947	12/1962	Cashmore, Newport

* The spelling was altered in November 1951 to read *County of Caernarvon*.

No 1012 *County of Denbigh* in the final GWR livery, and carrying a single chimney – only the first of the class, No 1000 *County of Middlesex*, was originally fitted with a double chimney. *Locomotive & General Railway Photographs*

valance was much narrower than in previous traditional practice, although its depth was increased slightly where the crosshead-driven pump was bolted on, just behind the cylinder on the right-hand side.

Construction

The Counties were built in two lots, Nos 354 and 358, between August 1945 and April 1947. Coming out with an axle load of 19 tons 14 cwt on each coupled axle, just under the 20 tons maximum, coupled with the use of two cylinders, their route availability was slightly more restricted than the Castles. The major cause of this was a more pronounced hammer blow on the track, which precluded their use on certain sections of LMS line, barring their use in particular on the Wolverhampton to Penzance routes. Initially, the 30 locomotives were to be numbered from 9900, but the final decision preferred numbers from 1000 upwards. The first, No 1000 *County of Middlesex*, followed immediately after the last of the LMS Class 8F 2-8-0s. Naming did not begin until 1946,

although the Counties already built had been turned out in pre-war livery, but without any lining below the footplate.

Tenders

With the locomotives came a new design of tender. The Hawksworth 4,000-gallon tender had the same basic dimensions as the previous Collett designs, with a fully laden weight of 49 tons, carrying seven tons of coal. The three-axle underframe with a 15ft 0in wheelbase divided equally, resulted in axle loads of 16 tons, 16 tons, and 17 tons, on 4ft 1½in wheels. It was a completely different design from previous GWR standard practice, with deep, continuous plate frames, stayed at 6ft 0in apart. The tanks and bunker formed an all-welded sheet steel assembly, and it was intended to be self-trimming.

From the operational viewpoint, footplate crews' opinions of the new tender were a little critical, perhaps because the layout was so different from the usual Swindon products. The sloping sides of the coal space were something of a disadvantage to shovelling coal forward, and unlike the older designs of tender the doors on the new flush-sided model opened out onto the footplate, and

not into the coal space. The traditional tool-boxes disappeared with the arrival of these tenders, with the fire irons kept in a tunnel on the right-hand side.

Operations

The Counties caused a stir when they first appeared, not least among the Great Western purists. Two diagrams were issued for the class, A16, and A17 for the double chimney locomotive, with power class D, and restriction to 'red' routes. The white X immediately below the route/power code disc on the cab side, signified that some overloading was permitted. Originally, No 1000 was power class E.

When new Nos 1000–1018 were turned out in pre-war GWR livery but without names, and while naming was resumed from March 1946, it was not until April 1948 that the last of the class was named. The remaining members of the class were given names when they were built.

In service, the Counties were first confined to the main lines between Paddington and Penzance, and Paddington to Wolverhampton. The first allocations included: Old Oak Common (London) 8, Bristol 7, Wolverhampton 5, Laira 3, Newton Abbot 2, Westbury 1, Exeter 1, Truro 1.

As BR locomotives the 30 Counties were

The last years of the GWR saw some dramatic changes, and Hawksworth's Counties could be counted among them. No 1027 *County of Stafford*, almost new, with single chimney and still sporting vestiges of GWR colours, is seen here shortly before nationalisation. *G. W. Sharpe*

The Counties were the most powerful two-cylinder design built by the GWR, and although more restricted in their route availability, they were widely travelled. No 1005 *County of Devon* is shown here in familiar territory, at Shrewsbury. *G. W. Sharpe*

rated as power class 6MT the same as the former LMS Royal Scot class three-cylinder locomotives which were 6P (later 7P). Work undertaken by the Counties in their BR days involved almost every type of train, on the main lines from Paddington to Bristol and the West Country, Wales, Chester, and the north. Their earliest BR allocations still found them at Old Oak Common, Bristol and Wolverhampton for the most part, but by 1953 redistribution had carried them further afield. In May 1953 the Counties were allocated as follows: Laira 5, Truro 2, Bristol 6, Neyland 4, Chester 4, Shrewsbury 5, Wolverhampton 4.

For the next few years further reallocations and changes of operational area re-sulted in the distribution of Counties to the following depots by 1960: 82B St Phillips Marsh 1000/5/9/11/14/21/24/27/28, 82C Swindon 1010/12/15/19/23, 83C Exeter St David's 1007, 83D Laira 1006/17, 83G Penzance 1002/3/4/8/18, 84G Shrewsbury 1013/16/22/26/26, 87H Neyland 1001/20/29.

Withdrawals began with Nos 1004 and 1026 in September 1962, when they were only 17 and 15 years old respectively. That same year saw them disappear from Devon and Cornwall. The Counties' final year of service was destined to be 1964, although with the number of already withdrawn only eight remained. Nos 1000/10/11/12/13/14/20/24, were all allocated to Swindon. Six were cut-up by Cashmore of Newport between December 1964 and March 1965, with the remaining two scrapped by Hayes of Bridgend and Swindon Works in July 1964. None survived for preservation.

CHAPTER SEVEN

THE LAST YEARS, AND PRESERVATION

Changing trends

The last few years of independence of the Great Western Railway were characterised by some significant changes, and in the light of an austere post-war economy, most developments were geared to general-purpose motive power. Under Hawksworth's guidance, just as in Churchward's day, motive power design and construction policy was undergoing radical change. Some of this had already been witnessed in the shape of the new mixed traffic County and Modified Hall classes, and a scheme was being developed to convert many steam locomotives to oil burning. Main line express passenger traffic was still in the hands of large steam types, but looming on the horizon were the prototype designs for gas turbine locomotives, entering a new era of rail traction technology. While the company's motive power policy was heavily influenced by North American practice in the early 1900s, it was not so in the late 1940s, when gas turbines were pre-ferred to the up-and-coming diesel-electric motive power. In retrospect, and bearing in mind the expensive experiment with diesel-hydraulic traction on BR Western Region, it may have been better to have looked towards America again, and developed the diesel-electric option earlier.

During this period, Hawksworth had introduced simplification to GWR design and construction practices, with plans for a number of other steam locomotives. The Modified Halls were considered a development of the most successful two-cylinder 4-6-0 design – the Hall class – but the very extensive nature of those changes could equally have been considered as a com-

Essentially, the same boiler – Standard No 1 – on different classes makes for interesting comparison of design policies between Hawksworth and Churchward/Collett. The Modified Hall is equally as typical of GWR tradition, as the Grange class on the right. *Lens of Sutton*

The 'de-luxe' model 43XX mogul turned out by Collett as 93XX series locomotives, with their additional weights at the front end, were altered by BR in the 1950s and renumbered in the 73XX series. No 7322 shown here was renumbered from 9300 in March 1957. *L. C. Jacks*

pletely new locomotive type. The most notable change in Hawksworth's day was the arrival in August 1945 of the radically different County class 4-6-0. It would have been even more revolutionary so far as Swindon was concerned if the scheme for outside Walschaerts valve gear and an LMS style tender had been used. Proposals for three other new steam locomotives were developed between 1944 and 1946, including a new lightweight 4-4-0 to replace the elderly Dukes and Bulldogs. In 1945, yet another proposal for a locomotive with outside valve gear appeared – a 2-6-0 pannier tank which included a selection of Swindon's standard components, from a 2301 class boiler to a Manor class smokebox.

The austerity economy of the late 1940s saw several schemes to improve the interchangeability of boilers, not only within the GWR, but extending beyond the company's boundaries. One particularly interesting idea included plans to equip a Hall class locomotive with an LMS Class 8F 2-8-0 boiler, of which Swindon Works had substantial experience during the war. Other ideas included schemes to provide the ROD 2-8-0s with Swindon's Standard No 1 boilers, and

perhaps most surprisingly a King class locomotive with a Southern Railway Merchant Navy 4-6-2 locomotive boiler. These developments foreshadowed a proposal for a new Pacific locomotive design – the second on the Great Western – which included an arrangement of cylinders and valve gear similar to that on the existing King class. The final proposal, made in 1946, was similar to the Stanier design for the Princess class on the LMS, particularly above the footplate, but with its four 16¼in × 28in cylinders, on 6ft 6in coupled wheels exerting a tractive effort identical with the King class 4-6-0s. From these projected designs only the 2-6-0 actually came into being, albeit in a very much modified form as the 1500 class 0-6-0 pannier tank, built in 1949. This was the last new locomotive design introduced by the GWR's last Chief Mechanical Engineer, who by that time had become chief mechanical engineer of the Western Region of British Railways.

Oil burners

With the fuel shortage of the immediate post-war period, the GWR had outlined a scheme to convert substantial areas of its system for operation by a fleet of oil-fired locomotives. The government of the day was in addition to removing certain oil fuel duties promising to offer financial incentives to those wishing to convert from coal to oil fuel. The Great Western, in company with the other three main line railway companies, had been authorised to convert no fewer than 1,217 locomotives.

Four classes of GWR locomotive were involved: sixty-three 28XX 2-8-0s, twenty-five Castle class 4-6-0s, eighty-four Hall class 4-6-0s, plus some 43XX 2-6-0s, providing a representative selection of freight, express passenger, and mixed-traffic locomotives. The ambitious scheme of both government and railway company was stifled in its early stages due to the inability to fund the large-scale import of fuel oil. The Great Western had by 1946 formulated plans to install fuelling equipment at Plymouth, Newton Abbot, Llanelly, Landore, Cardiff, Newport, Severn Tunnel Junction, Bristol, Gloucester, Westbury, Swindon, Didcot, Reading, and Old Oak Common.

Stationary and
Controlled Road Tests

Two members of the Modified Hall class were the subject of scrutiny in the early BR years. No 6990 *Witherslack Hall* took part in the 1948 Interchange Trials, and No 7916 *Mobberley Hall* underwent exhaustive stationary and controlled road tests. The examination of and results from the testing of the latter were instrumental in improving the performance of this design. An important aspect of the tests, the findings from which were published in 1951, was the establishment of national standards for locomotive testing. At Swindon, under the guidance of S. O. Ell, the idea that the performance of locomotives should be related to constant steaming, was developed to produce the extensive testing programmes undertaken by British Railways in the 1950s. The principle established during those early years was that the thermodynamic performance of a locomotive at a constant speed with a predetermined rate of evaporation on the test plant, could be reproduced out on the road, with the same evaporation rate at varying speeds. With No 7916 *Mobberley Hall* on test the blastpipe diameter was reduced to 5⅛in, and the internal taper of the chimney to 1 in 14, with the jumper ring removed. The distance from the top of the blastpipe to the chimney choke was increased to 2ft 4in, in order to provide more efficient ejection.

Changes in the degree of superheat,

already introduced during Hawksworth's day, along with other changes in traditional GWR locomotive design, were supplemented by the draughting changes made following these tests. Testing during BR days used Blidworth Grade 2B coal as the standard fuel, having a lower calorific value than the Welsh coals used by the GWR. It was suggested that the design of the firebox in Swindon's locomotives and the use of the highest quality coals were factors which weighed heavily against the Hall 4-6-0, and its poor performances in the 1948 Interchange Trials. But when No 7916 *Mobberley Hall*, new in April 1960, came to be tested with the attendant modifications to draughting, the results largely dispelled the myth that GWR locomotives would only produce their best performances with the best coals. Traditional design practices meant that the Standard No 1 boiler was not being used to its full potential, and the elimination of the jumper top blastpipe, and dimensional changes produced a marked improvement.

The controlled road tests were carried out between Wantage Road and Stoke Gifford Yard, and were the first to be conducted on British Railways with the steaming rate rigidly controlled. At the normal operating speeds for a mixed traffic type, the modified locomotive was producing its optimum drawbar horsepower between 20mph and 50mph, a figure only marginally improved on using the Welsh coals. However, at steam rates of 16,000 lb/hour and over, coal consumption of the lower grade fuels increased more rapidly than it did when burning the South Wales varieties.

An interesting if surprising result from the exhaustive tests carried out on the Halls was the poor efficiency – only 49% – recorded when using the Blidworth coals. This, considering the suggestions that Churchward, and GWR type boilers were amongst the best steamers, does not stand up to detailed comparison with other locomotive types. While the boiler efficiency at the front end limit was poor under maximum sustained conditions, the boiler efficiency increased to 74%, the highest for all locomotives tested. That the draughting alterations carried out in the 1950s were an improvement is beyond question, and the removal of jumper tops on the blastpipes together with other modifications was carried out on a number of ex-GWR types, including the Grange and Manor classes.

The major change to the Manors began with the testing of one of the new British Railways Standard Class 4MT 4-6-0s in 1951, which had similar tractive effort, and a high route availability but was far superior in performance. The boiler on the Standard Class 4 proved to be a better steam producer proportionally than that of the Modified Hall No 7916, tested at Swindon. The following year, 1952, No 7818 *Granville Manor* was tested. While it was widely recognised the Manors were poor steamers, quite how poor they were was emphasised when a maximum sustained steam rate of only 10,000 lb/hour was recorded. This figure was achieved with the traditional front end layout, with a jumper top blastpipe, and a nozzle diameter of 5⅛in, with a rather narrow chimney choke diameter of 13½in.

From its original condition, the initial alterations made to No 7818 for further test-

ing were: a reduction in blast nozzle diameter to 4¾in, the removal of the jumper ring, and provision of a temporary stovepipe chimney with similar proportions to that fitted on the BR Standard 4-6-0. Further improvement resulted from a reduction in the blast nozzle diameter to 4⅝in, and increasing the amount of free air space through the grate by removing sections of firebars. With these changes, the improvement in the Manors' performance was little short of phenomenal, and on test an evaporation rate of no less than 20,400 lb/hour was sustained.

Other comparisons with the BR Standard Class 4 showed the improved Manors to be more economical than their BR Standard counterparts. In terms of indicated horsepower (IHP) the improved performance of the Manor class outstripped even that of the Hall class, when both locomotives were working up to the evaporative limit of the boiler, in respect of coal consumption of both types.

FIRING RATES – MODIFIED HALL CLASS
7990 MOBBERLEY HALL

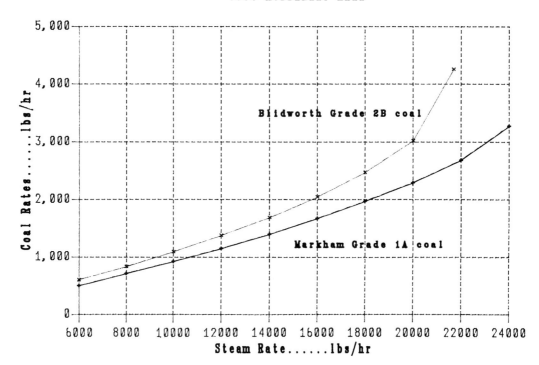

INDICATED HP/DRAWBAR HP
MANOR CLASS 4-6-0

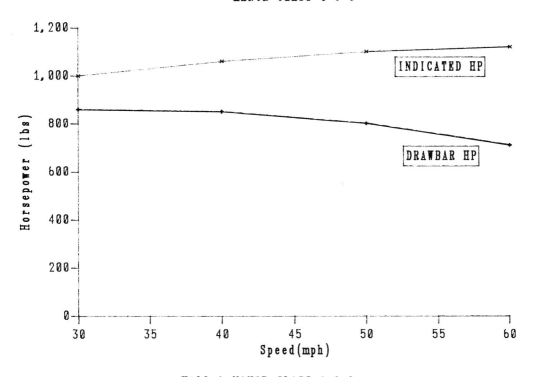

HALL & MANOR CLASS 4-6-0
IHP Using Blidworth Coal
At 50 mph

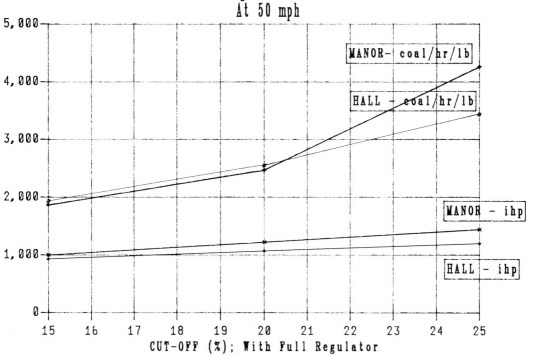

Hall class No 5970 *Hengrave Hall*, stands at Shrewsbury, complete with Hawksworth 4,000-gallon tender and later BR livery of lined green. The white X below the cabside route disc signifies that some overloading with this locomotive was permitted. *G. W. Sharpe*

Livery Styles

The Great Western two-cylinder classes, including both 2-6-0 and 4-6-0 designs, saw three distinct shades of green between the turn of the century and nationalisation. The major changes after 1948 introduced two very different painting styles, with the final schemes adopting lined green livery, in common with most ex-GWR types running on the Western Region.

Predominant features on locomotives around the turn of the century emanated from a livery change introduced by William Dean in 1881, and locomotive green was enhanced with lining in two different styles. Dean's pioneer 4-6-0 No 36 had boiler bands picked out in black, separated from the green by ⅛in orange lines. The Indian Red outside frames were also lined, while the tender sides had three lined panels, with the centre panel carrying the company's monogram. About 1900 this was changed and tenders were lined in a single panel on each side. Motion and numerous other details were bright finished, with the polished brass

domes, safety valve covers, and cab spectacle plate window frames.

From 1906 to 1947 the standard livery adopted by the Great Western was lined green for passenger locomotives, although the shade became lighter in 1928. The lining applied to boiler bands was again a black line flanked by twin ⅛in wide orange lines. The same thin orange line highlighted the footplate valances and footsteps, with twin orange lines forming a panel on the cylinder wrappers. All paintwork below, and including the upper surfaces of the running boards was black. Wheel splashers sported a 2in wide black border, with a ⅛in wide orange line inside.

Number plates cast in brass or iron with figures 5¾in high and carried on the cabside, were a feature retained in BR days when the system-wide numbering scheme for locomotives was introduced. The brass letters on nameplates were a standard 3½in high while small letters, such as 'of' in *Lady of Lynn* were 2in high. The letters and surrounding brass beading were rivetted to a steel arc shape plate.

In 1904, the gartered arms of London and Bristol began to appear on tender sides, flanked by the words 'Great Western', replacing the earlier company monogram. In the late 1920s, the style was changed again, with the company name flanking the crests of London and Bristol in a new coat of arms. The 'shirt button' GWR monogram became fashionable in 1934, but this was superseded in 1942 for passenger locomotives by the coat of arms insignia flanked by the initials, GW – lesser classes sported GWR in the same style lettering, without the coat of arms.

The two periods of war, 1914–1918 and 1939–1945, produced some interesting variations, and after World War I it became policy to apply lining to express locomotives only. This included the Saints but was later extended to cover the Hall and subsequent mixed traffic classes. World War II saw unlined black livery applied to the two-cylinder types, while the first locomotive to be repainted in fully lined out pre-war style GWR green livery was the new County class 4-6-0 No 1000.

After nationalisation all the two-cylinder 4-6-0s including the Saints were destined to be repainted in BR mixed traffic black, lined out in red, cream and grey, in the first standard liveries of 1949. But even before this during an experimental period another 'first'

was with No 4946 *Moseley Hall* becoming the first steam locomotive from any company to show any outward sign of a change of ownership when it appeared in January 1948 in fully lined GWR livery but carrying the words British Railways on the tender in GWR style lettering. With the exception of those locomotives which carried the W prefix to the running number, no GWR two-cylinder tender engine carried an experimental BR livery. The BR black, lined red, cream and grey, with lion-and-wheel insignia, was applied to Saint, Hall, Modified Hall, Grange, Manor, and County 4-6-0s, and the 43XX 2-6-0s. The 26XX moguls were soon to disappear in BR days, and were not repainted.

During the 1950s changes in BR organisation were among influences that allowed greater autonomy to the regions, which made itself felt in the altered colours on locomotives and rolling stock. In 1955 No 6997 *Bryn-Ivor Hall* was outshopped from Swindon wearing lined green livery, of almost the

Another livery change is seen on No 6341, just off Swindon Works after a general overhaul, this time in plain green livery, in July 1960. *L. C. Jacks*

Mogul No 7313, fresh from Wolverhampton Stafford Road Works, is in fully lined passenger green livery in July 1958, at Tyseley shed. This view clearly shows the fixing of the front frame tie rod supports, and pony truck spring housing on the front running plate. *L. C. Jacks*

same style as that used by the Great Western Railway. This marked the beginning of a move away from lined black livery for mixed traffic types, and involved tank as well as tender designs. Some 43XX moguls appeared in unlined green, just as they had many years previously, but this time carrying BR's new heraldic device, while others received full lining. After 1956 lined green livery with the new BR crest could be found on Hall, Modified Hall, County, Grange, Manor, and 43XX moguls.

The dimensions of lining, and size and position of livery areas were the same in the 1960s as had been laid down in 1949, using both synthetic and oleo-resinous paints. On green locomotives, the livery colour was applied to boiler barrel, firebox, cab sides and spectacle plate, tender sides and rear, valances and outer faces of wheel splashers. Black was applied to cab roofs, upper surfaces of running boards, front frame extensions, and all chassis details. Orange lining was separated by a black line on boiler and firebox lagging bands, wheel splashers had a black border, with ⅛in orange line inside, and twin orange lines on the cylinder wrappers. Cab and tender sides (not the rear panel) were lined with a 1in black line, flanked by ⅛in orange line on either side. Locomotives treated to mixed traffic black livery in the earlier style had red, cream and grey lining, with twin red lines forming a panel on the cylinder wrapper plates.

Preservation and survival

Some 24 former GWR two-cylinder locomotives have already been acquired for preservation in one form or another, with 4953 *Pitchford Hall* recently obtained (at the time of writing) for the Dean Forest Railway. None of the Counties, Saints, Granges, or Aberdares was rescued, although there is an ambitious plan to reconstruct No 4942 *Maindy Hall* as a replica Saint. This imaginative idea put forward by the Great Western Society would reverse the process initiated in December 1924, when Saint class 4-6-0 2925 *Saint Martin* was rebuilt to become the forerunner of the Hall class.

Out of the current total of 24 preserved two-cylinder types there are two 43XX moguls, eight Halls, five Modified Halls, and nine Manors. It is perhaps not so surprising that so many Manors have survived,

Hall class versatility as a mixed traffic type knew practically no bounds. No 4963 is passing mogul No 5319, hauling a lengthy goods train. *G. W. Sharpe*

Not all the Hawksworth self-trimming tenders were paired with the newer locomotives. Here Hall class No 4951 has the straight-sided 4,000-gallon tender attached, sporting early BR lined black colours with 'lion-and-wheel' emblem. *G. W. Sharpe*

Grange class locomotives, perhaps the most popular of
the two-cylinder types, also saw service on the most
prestigious passenger trains. In BR days No 6818
Hardwick Grange is seen at Cardiff General at the head of
the up Red Dragon. *Historical Model Railway Society/
R. E. Lacey*

since their light weight, and much improved
performance render them ideal for operation
on the many preserved lines around the
country. Three of their number can be
found on the Severn Valley Railway, two
more on the Gloucestershire & Warwick-
shire Railway, and one each for the Gwili
Railway, Dart Valley, the Cambrian Railway Society at Oswestry, and the Great
Western Society's Didcot Railway Centre.
The Halls and Modified Halls show a similarly widespread distribution, while the
remaining moguls can be found at Didcot
and on the Severn Valley Railway.

Locations of the former Great Western
two-cylinder tender locomotives, many of
which are now beyond their 50tth birthdays,
at the time of going to press were:

43XX Moguls

5322   Didcot Railway Centre
7325   Severn Valley Railway (formerly No 9303)

Hall Class 4-6-0

4920	*Dumbleton Hall*	Dart Valley Railway
4930	*Hagley Hall*	Severn Valley Railway
4936	*Kinlet Hall*	Peak Railway, Matlock
4942	*Maindy Hall*	Didcot Railway Centre
4983	*Albert Hall*	Birmingham Railway Museum Tyseley
5900	*Hinderton Hall*	Didcot Railway Centre
5952	*Cogan Hall*	Gloucestershire & Warwickshire Railway
5972	*Olton Hall*	Procor (UK) Ltd, Horbury Jct, Wakefield

Modified Hall Class 4-6-0

6960	*Raveningham Hall*	Severn Valley Railway
6989	*Wightwick Hall*	Quainton Railway Centre, Aylesbury
6990	*Witherslack Hall*	Great Central Railway, Loughborough
6998	*Burton Agnes Hall*	Didcot Railway Centre
7903	*Foremarke Hall*	Swindon & Cricklade Railway

Manor Class 4-6-0

7802	*Bradley Manor*	Severn Valley Railway

7808	Cookham Manor	Didcot Railway Centre
7812	Erlestoke Manor	Severn Valley Railway
7819	Hinton Manor	Severn Valley Railway
7820	Dinmore Manor	Gwili Railway, Carmarthen
7821	Ditcheat Manor	Gloucestershire & Warwickshire Railway
7822	Foxcote Manor	Cambrian Railway Society, Oswestry
7827	Lydham Manor	Dart Valley Railway
7828	Odney Manor	Gloucestershire & Warwickshire Railway

From the viewpoint of sheer numerical strength, the former GWR two-cylinder classes are popular with preservationists and the privately operated steam railways as they were during their years of service. It would perhaps be too much to ask if some enterprising organisation could resurrect a Grange, or even a County in the same way as plans to provide a replica of the progenitor, the Churchward Saint class 4-6-0.

Preserved Hall class 4-6-0 No 6998 Burton Agnes Hall on a special centenary working on the Maidenhead–Bourne End branch on 15 July 1973. Stanley Creer

APPENDIX

Lot Nos, Works Nos, and Running Nos summary

Lot Nos	Works Nos	Running Nos	Wheel arrangement	Built
106	1551	36	4-6-0	1896
116	1723	2601	4-6-0	1899
	1724	2602	2-6-0	1901
	1725–1732	2603–2610	2-6-0	1903
123	1796–1805	2611–2620	2-6-0	1903
128	1886	33	2-6-0	1900
131	1908–1927	2621–2640	2-6-0	1901
132	1928	100	4-6-0	1902
133	1929–1941	2641–2653	2-6-0	1901
	1942–1948	2654–2660	2-6-0	1902
135	1949–1968	2661–2680	2-6-0	1902
138	1990	98	4-6-0	1903
145	2024	171	4-6-0*	1903
154	2106	172	4-4-2‡	1905
	2107–2112	173–178	4-6-0	1905
	2113–2114	179–180	4-4-2‡	1905
156	2125	2601	2-6-0	1906
158	2128–2137	181–190	4-4-2‡	1905
164	2199–2208	2901–2910	4-6-0	1906
170	2259–2278	2911–2930	4-6-0	1907
183	2396–2415	4301–4320	2-6-0	1911
185	2426–2435	2931–2940	4-6-0	1911
189	2476–2485	2941–2950	4-6-0	1912
192	2506–2510	2951–2955	4-6-0	1913
193	2516–2525	4321–4330	2-6-0	1913
194	2526–2535	4331–4340	2-6-0	1913
198	2552	4341	2-6-0	1913
	2553	4342	2-6-0	1913
	2554–2571	4343–4360	2-6-0	1914
202	2612–2631	4361–4380	2-6-0	1915
204	2652–2671	4381–4399	2-6-0	1916
		4300	2-6-0	1916
205	2672–2676	5300–5304	2-6-0	1916
	2677–2681	5305–5309	2-6-0	1917
206	2682–2701	5310–5329	2-6-0	1917
207	2702–2709	5330–5337	2-6-0	1917
	2710–2721	5338–5349	2-6-0	1918
208	2722–2731	5350–5359	2-6-0	1918
	2732–2741	5360–5369	2-6-0	1919
209	2742–2751	5370–5379	2-6-0	1919
	2752–2761	5380–5389	2-6-0	1920
211	2790–2806	5390–5399	2-6-0	1920
		6300–6306	2-6-0	1920
	2807–2817	6307–6317	2-6-0	1921
212	2818–2841	6318–6341	2-6-0	1921
216	2887–2906	6342–6361	2-6-0	1923
	(Works Nos were not issued after Lot No 217)			
218	—	6370–6399	2-6-0	1921
		7300–7302	2-6-0	1921
		7303–7304	2-6-0	1922
222	—	7305–7318	2-6-0	1921
		7319	2-6-0	1922
230	—	7320–7321	2-6-0	1925
		6362–6369	2-6-0	1925
254	—	4901–4905	4-6-0	1928
		4906–4970	4-6-0	1929
		4971–4980	4-6-0	1930

Lot Nos	Works Nos	Running Nos	Wheel arrangement	Built
268	—	4981–4999	4-6-0	1931
		5900	4-6-0	1931
275	—	5901–5920	4-6-0	1931
276	—	9300–9319§	2-6-0	1932
281	—	5921–5940	4-6-0	1933
290	—	5941–5950	4-6-0	1935
297	—	5951–5957	4-6-0	1935
		5958–5965	4-6-0	1936
304	—	5966–5975	4-6-0	1937
308	—	6800–6819	4-6-0	1936
		6820–6859	4-6-0	1937
		6860–6879	4-6-0	1939
311	—	5976–5985	4-6-0	1938
316	—	7800–7811	4-6-0	1938
		7812–7819	4-6-0	1939
327	—	5986–5994	4-6-0	1939
		5995	4-6-0	1940
333	—	5996–5999	4-6-0	1940
		6900–6905	4-6-0	1940
338	—	6906–6910	4-6-0	1940
		6911–6915	4-6-0	1941
340	—	6916–6935	4-6-0	1941
		6936–6950	4-6-0	1942
		6951–6958	4-6-0	1943
350	—	6959–6970	4-6-0	1944
354	—	1000–1009	4-6-0	1945
		1010–1019	4-6-0	1946
358	—	1020–1022	4-6-0	1946
		1023–1029	4-6-0	1947
366	—	6971–6980	4-6-0	1947
		6981–6990	4-6-0	1948
368	—	6991–6995	4-6-0	1948
		6996–6999	4-6-0	1949
		7900–7906	4-6-0	1949
		7907–7919	4-6-0	1950
376	—	7920–7929	4-6-0	1950
377	—	7820–7829	4-6-0	1950

* No 171 ran as a 4-4-2 from October 1904 to July 1907.

‡ Nos 172, 179-190 built as 4-4-2s and converted to 4-6-0s between April 1912 and January 1913.

§ Renumbered 7322–7341 after modification, 1956–1959

INDEX